GRAPHS, GROUPS
AND SURFACES

For Liz

NORTH-HOLLAND
MATHEMATICS STUDIES 8

Graphs, Groups and Surfaces

ARTHUR T. WHITE

Western Michigan University
Kalamazoo, Mich.
U.S.A.

1973

NORTH-HOLLAND PUBLISHING COMPANY – AMSTERDAM ● LONDON
AMERICAN ELSEVIER PUBLISHING COMPANY, INC. – NEW YORK

Library of Congress Catalog Number: 73-86088
ISBN North-Holland:
Series: 0 7204 2600 6
Volume: 0 7204 2608 1
ISBN American Elsevier: 0 444 10570 0

PUBLISHERS:

NORTH-HOLLAND PUBLISHING COMPANY – AMSTERDAM
NORTH-HOLLAND PUBLISHING COMPANY, LTD.–LONDON

SOLE DISTRIBUTORS FOR THE U.S.A. AND CANADA:

AMERICAN ELSEVIER PUBLISHING COMPANY, INC.
52 VANDERBILT AVENUE
NEW YORK, N.Y. 10017

PRINTED IN THE NETHERLANDS

FOREWORD

This text is based on a course that I taught at Western Michigan University in 1971. The material is suitable for presentation at the graduate or advanced undergraduate level, and assumes only an introductory knowledge of group theory and of point-set topology. One of the features of the subject matter is that many results of the general theory can be readily visualized, or modeled, so that the student is constantly reassured that what he is doing has meaning, is more than a formal manipulation of symbols.

The focus of attention is an interaction among graphs, groups, and surfaces (see Figure 0-1). After a brief introduction to the theory of graphs, the relationship between graphs and groups will be explored. For every graph there is an associated group, called the automorphism group of the graph. Conversely, for every group presentation, there is an associated graph, called a Cayley color graph of the group. Both associations will be studied, with emphasis on the latter. An excursion into combinatorial topology will follow; Euler's generalized polyhedral formula and the classification theorem for closed 2-manifolds will be discussed. Imbedding problems in graph theory will be examined at some length (here a graph and a surface come together, often with the aid of a group); it will be seen that the famous four-color conjecture can be stated in this context. The five-color theorem will be established. The Heawood map-coloring theorem will then be studied in detail. (This theorem determines a chromatic number for every closed 2-manifold except the sphere -- a truly astounding result!) The theory of quotient graphs and quotient manifolds (and quotient groups) is at the heart of the proof of this theorem (as well as many others) and will be presented; this theory relies upon each of the concepts mentioned in this paragraph and serves as the unifying feature of this text.

Several peripheral (but significant) results are stated without proof. An effort has been made to provide those proofs of theorems which are most indicative of the charm and beauty of the subject and which illustrate the techniques employed. Proofs missing in the text can be supplied by the reader, as part of the problem sets (problems marked "*" are difficult; those marked "**" are as yet unsolved), or can be found in the references. A bibliography is provided, for further reading; items (c) or (g), (d), and (e) respectively are particularly suitable for more extensive treatments of the theories of graphs, groups, and surfaces, which are seen interacting in this text.

I thank Mrs. Darlene Lard, for typing the manuscript, and Mr. Paul E. Himelwright, for drawing the figures.   A.T.W.

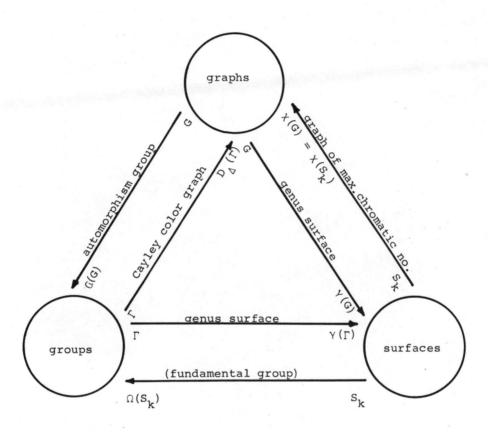

Figure 0-1

TABLE OF CONTENTS

Chapter 1: Historical Setting

In coloring the regions of a map, one must take care to color
differently any two countries sharing a common boundary line, so that
the two countries can be distinguished. One would think that an
economy-minded map-maker would wish to minimize the number of colors
to be used for a given map, although there appears to be no historical evidence of any such effort. Nevertheless a conjecture was made,
over twelve decades ago, to the effect that four colors would always
suffice for a map drawn on the sphere, the regions of which were all
connected. The first reported mention of this problem (see [4] and
[47]) was by Francis Guthrie, through his brother Frederick and
Augustus de Morgan, in 1852. The first written references were by
Cayley, in 1878 and 1879. Incorrect "proofs" of the Four Color Conjecture were published soon after by Kempe and Tait. The error in
Kempe's "proof" was found by Heawood [29] in 1890; this error has
reappeared in various guises in subsequent years. Ore and Stemple
[48] have shown that any counterexample to the conjecture must
involve a map of at least 40 regions. The conjecture continues to
provide one of the most famous unsolved problems in mathematics.

It is an astonishing fact that several related, seemingly much
more difficult, map-coloring problems have been completely solved.
Chief among these is the Heawood Map-coloring Conjecture, which gives
the chromatic number for _every_ closed 2-manifold _other_ _than_ _the_
sphere; we state the orientable case:

$$\chi(S_k) = f(k) = \left\lceil \frac{7 + \sqrt{1 + 48k}}{2} \right\rceil, \text{ for } k > 0 ,$$

where k is the genus of the closed orientable 2-manifold S_k.
Heawood showed in 1890 [29] that $\chi(S_k) \leq f(k)$, and in 1891 Heffter
[30] showed the reverse inequality for a possibly infinite set of
natural numbers k; almost eight decades passed before it was shown
that $\chi(S_k) \geq f(k)$, for all $k > 0$. In 1965 this problem was given
the place of honor on the dust jacket for Tietze's _Famous_ _Problems_
_of_ _Mathematics_ [67]. An outline of the major portion of the solution now follows.

The dual of a map drawn on S_k is a pseudograph imbedded in S_k, and it can be shown (see Section 8-4) that $\chi(S_k) \geq f(k)$, for $k > 0$, provided the complete graph K_n has genus given by

$$\gamma(K_n) = \left\{ \frac{(n-3)(n-4)}{12} \right\}, \ n \geq 7 . \qquad (*)$$

Heawood established (*) for n = 7 in 1890, and Heffter for $8 \leq n \leq 12$ in 1891; Ringel handled n = 13 in 1952. The first major breakthrough occurred in 1954, when Ringel showed (*) for $n \equiv 5 \pmod{12}$. During 1961-1965, Ringel treated the residue cases 7, 10, and 3 (mod 12), while independently Gustin settled the cases 3, 4, and 7. Gustin's method involved the powerful and beautiful idea of quotient graph and quotient manifold, and relies upon the fact that K_n can be regarded as a Cayley color graph for a group presentation; thus graph theory, group theory, and surface topology are combining to solve this famous problem of mathematics.

In 1965, Terry, Welch, and Youngs announced their solution to case 0. Gustin, Ringel, and Youngs finished the remaining residue cases (mod 12), except for the isolated values n = 18, 20, and 23; their work was announced in 1968 [56]. In 1969, Jean Mayer (a Professor of French Literature) [43] eliminated the last three obstinate graphs by ad hoc techniques.

Much of the work of Ringel, Terry, Welch, and Youngs was made possible by Gustin's theory of quotient graphs and quotient mani- folds; this theory was developed and modified by Youngs, who also introduced the theory of vortices [75]. The theory is considerably more general than was needed to prove the Heawood Map-coloring Theorem, and has been unified and developed in its full generality by Jacques [32], in 1969. Jacques' results will be presented in Chapter 9, together with many applications to other imbedding problems in graph theory. This will be the focal point of the text, and it illustrates vividly the fruitful interaction among graphs, groups, and surfaces.

The conjunction of graph theory, group theory, and surface topology described above is foreshadowed, in this text, by several pairwise interactions among these three disciplines. The Heawood Map-coloring Theorem is proved by finding, for each surface, a graph of largest chromatic number that can be drawn on that surface. Equivalently (as it turns out) we find, for each complete graph, the surface of smallest genus in which it can be drawn. The extension of this latter problem to arbitrary graphs is natural; the solution is

particularly elegant for graphs which are the Cayley color graphs of
a group. We are led in turn to the problem of finding, for a given
group, a surface of minimum genus which represents the group in some
way.

Dyck [20] (see also Burnside [10], Chapters 18 and 19) considered maps, on surfaces, that are transformed into themselves in
accordance with the fixed group Γ, acting transitively on the
regions of the map. Any such map gives an upper bound for the parameter $\gamma(\Gamma)$ discussed in Chapter 7 of this text, as a "dual" formed
in terms of Burnside's white regions gives a Cayley color graph for
Γ.   (Cayley [11] defined his color graphs as complete symmetric
digraphs, corresponding to the choice Γ less the identity element
as a generating set for Γ; it is sensible to extend his definition
to any generating set for the group in question.) Brahana [9]
studied groups represented by regular maps on surfaces; these maps
correspond to presentations on two generators, one of which is of
order two. In this context the group acts transitively on the edges
of the map, and again an upper bound for $\gamma(\Gamma)$ is obtained. In
Chapter 7, we regard Γ as acting transitively on the vertices of
the map induced by imbedding a Cayley color graph $D_\Delta(\Gamma)$ for Γ in
a surface; in Chapter 4, we show that the automorphism group of
$D_\Delta(\Gamma)$ is isomorphic to Γ, independent of the generating set Δ
selected for Γ, so that in this sense $D_\Delta(\Gamma)$ provides a "picture"
of Γ.   But more: many properties of Γ, such as commutivity,
normality of certain subgroups, the entire multiplication table, can
be "seen" from the picture provided by $D_\Delta(\Gamma)$.   Thus it is natural
to seek the simplest surface on which to draw this picture; this is
given by the parameter $\gamma(\Gamma)$.

This point of view may give a surface of lower genus for a
given group than the other two approaches listed above; for example,
the group $\Gamma = Z_2 \times Z_4$ is toroidal for Dyck (or Burnside) and for
Brahana, yet $\gamma(Z_2 \times Z_4) = 0$.

There is one correspondence depicted in Figure 0-1 which we do
not discuss in this text: to every surface S_k there corresponds a
unique group, $\Omega(S_k)$, called the fundamental group of the surface;
the groups $\Omega(S_k)$ have been completely determined -- they are given
by 2k generators $a_1, b_1, \ldots, a_k, b_k$ and the single defining
relation $a_1 b_1 a_1^{-1} b_1^{-1} \ldots a_k b_k a_k^{-1} b_k^{-1} = e$ (see, for example, [62].)
Each of the other five correspondences illustrated in Figure 0-1
(where the inner triangle commutes, for proper choice of Δ) is
germane, as outlined above, to the conjunction of graph theory,

group theory, and surface topology described in this introduction
and which we now begin to develop.

Chapter 2:   A Brief Introduction to Graph Theory

In this chapter we introduce basic terminology from the theory of graphs that will be used in this text. We will give several binary operations on graphs; these will enable us to construct more complicated graphs, and hence to build up our store of examples of frequently encountered graphs.

We emphasize that the material introduced here is primarily for the purpose of later use in this text; for a considerably more thorough introduction to graph theory, see [4] or [28].

2-1. Definition of a Graph

Def. 2-1.   A graph G consists of a finite non-empty set V(G) of vertices together with a set E(G) of unordered pairs of distinct vertices, called edges. If x = [u,v] ∈ E(G), for u, v ∈ V(G), we say that u and v are adjacent vertices, and that vertex u and edge x are incident with each other, as are v and x. We also say that the edges [u,v] and [u,w], w ≠ v, are adjacent. The degree, d(v), of a vertex v is the number of edges with which v is incident. (Equivalently, d(v) is the number of vertices to which v is adjacent; i.e.

$$d(v) = |\{u \in V(G)|\ [u,v] \in E(G)\}|.)$$

If the vertices of G are labeled, G is said to be a labeled graph.

As a matter of notation, we usually write uv for [u,v]; p = |V(G)|; q = |E(G)|. The order of G is given by p.

Example: Let   G   be defined by:

$$V(G) = \{v_1, v_2, v_3, v_4\}$$

$$E(G) = \{v_1v_2, v_2v_3, v_3v_1, v_1v_4\};$$

then   G   may be represented by either Figure 2-1a or 2-1b, where the latter representation is more accurate, in a sense we will describe in Chapter 6.

G: G:

(a)   (b)
Figure 2-1.

Note: A <u>graph</u> may be more briefly defined as a finite one-dimensional simplicial complex.

Thm. 2-2.   For any graph   G, $\sum\limits_{i=1}^{p} d(v_i) = 2q.$

Proof: In summing the degrees, each edge is counted exactly twice.

Cor. 2-3.   In any graph   G,   the number of vertices of odd degree is even.

2-2.   Variations of Graphs

Def. 2-4.   A <u>loop</u> is an edge of the form   vv.   A <u>multiple edge</u> is an edge that appears more than once in   E(G).   A <u>directed edge</u> is an ordered pair of distinct vertices.   A <u>multigraph</u> allows multiple edges.   A <u>pseudograph</u> allows loops and multiple edges.   A <u>directed graph</u> (<u>digraph</u>) has every edge directed.   An <u>infinite graph</u> has infinite vertex set.

For example, see Figure 2.2.

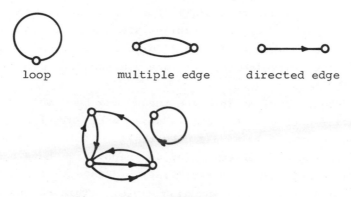

loop   multiple edge   directed edge

a directed pseudograph

Figure 2-2.

    The term "graph," unless qualified appropriately, disallows any
and all of the above variations.

                2-3.   Additional Definitions

Def. 2-5. A graph H is said to be a <u>subgraph</u> of a graph G if
           $V(H) \subseteq V(G)$ and $E(H) \subseteq E(G)$. If $V(H) = V(G)$, H is
           called a <u>spanning subgraph</u>. For any $\phi \neq S \subseteq V(G)$, the
           <u>induced subgraph</u> $\langle S \rangle$ is the maximal subgraph of G with
           vertex set S.

<u>Notation</u>: For $v \in V(G)$, $G - v$ denotes $\langle V(G) - v \rangle$. For $x \in E(G)$,
           $V(G - x) = V(G)$, and $E(G - x) = E(G) - x$.

    Certain subgraphs are given special names. We indicate these by
a series of definitions.

Def. 2-6. A <u>walk</u> of a graph G is an alternating sequence of ver-
           tices and edges $v_0, x_1, v_1, \ldots, v_{n-1}, x_n, v_n$ (or,
           briefly: $v_0, v_1, \ldots, v_{n-1}, v_n$) beginning and ending
           with vertices, in which each edge is incident with the

two vertices immediately preceding and following it.   (n
is the <u>length</u> of the walk.)   If $v_0 = v_n$,   the walk is
said to be <u>closed</u>; it is said to be <u>open</u> otherwise.   The
walk is called a <u>trail</u> if all its edges are distinct, and
a <u>path</u> if all the vertices are distinct.   A <u>cycle</u> is a
closed walk with $n \geq 3$ distinct vertices (i.e. $v_0 = v_n$,
but otherwise the v_i are distinct).

Two famous problems in graph theory may be described in terms of
the above definitions.   A graph is said to be <u>eulerian</u> if the graph
itself can be expressed as a closed trail.   (This corresponds to the
"highway inspector" problem; eulerian graphs have been completely and
simply characterized: see Harary [28], p. 64-65.)   A graph is said to
be <u>hamiltonian</u> if it has a spanning cycle.   (This corresponds to the
"traveling salesman" problem; hamiltonian graphs have <u>not</u> been completely characterized.   See Harary, p. 65-69, for some partial
results.)

<u>Def. 2-7</u>.   A graph   G   is <u>connected</u> if $u, v \in V(G)$ implies there
           exists a path in   G   joining   u   to   v.   A <u>component</u> of   G
           is a maximal connected subgraph of   G.

<u>Def. 2-8</u>.   The <u>distance</u>,   d(u,v),   between two vertices   u   and   v
           of   G   is the length of a shortest path joining them if
           such exists; if not, $d(u,v) = \infty$.

<u>Thm. 2-9</u>.   A connected graph may be regarded as a (finite) metric
           space.

           <u>Proof</u>:   See Problem 2-7.

For a partial converse to the above theorem, see Chartrand and
Kay [12].   By Theorem 2-9, every connected graph may be regarded as
a topological space.   (Actually, since the metric induces the discrete topology, we knew this already.)   In Chapter 6 we will see that
this is true in another sense also; that is every graph may be regarded as a subspace of E^3,   with all edges represented as <u>straight</u>
lines.   If we consider   G   as a topological space in this latter
sense, then   G   is connected as a graph if and only if it is connected as a topological space (see Problem 2-8).   The term "component"
is easily seen to mean the same in both contexts.   Furthermore, a

graph (as a subspace of E^3) is connected if and only if it is path connected; (see Problem 2.9.)

<u>Def. 2-10.</u> Two graphs G_1 and G_2 are said to be <u>isomorphic</u>
($G_1 \cong G_2$, or $G_1 = G_2$) if there exists a one-to-one,
onto map $\theta : V(G_1) \rightarrow V(G_2)$ preserving adjacency; that is,
$uv \in E(G_1)$ if and only if $\theta(u)\theta(v) \in E(G_2)$.

<u>Note</u>:   Isomorphism is an equivalence relation on the set of all graphs.

<u>Notation</u>: $\delta(G) = \min\{d(v) \mid v \in V(G)\}$.

$\Delta(G) = \max\{d(v) \mid v \in V(G)\}$.

<u>Def. 2-11.</u> If $\delta(G) = \Delta(G) = r$, we say that G is <u>regular</u> of degree r. (If $r = 3$, G is said to be <u>cubic</u>.)

<u>Thm. 2-12.</u> Let $\theta : V(G_1) \rightarrow V(G_2)$ give $G_1 \cong G_2$; then $d(\theta(v)) = d(v)$, for all $v \in V(G_1)$.

<u>Proof</u>:   See Problem 2-10.

<u>Cor. 2-13.</u> If G_1 is regular of degree r and $G_1 \cong G_2$, then G_2 is regular of degree r.

<u>Cor. 2-14.</u> Let the vertices of a graph G_1 have degrees $d_1 \leq d_2 \leq \cdots \leq d_n$, and the vertices of a graph G_2 have degrees $c_1 \leq c_2 \leq \cdots \leq c_n$. If $d_i \neq c_i$, for some $1 \leq i \leq n$, then G_1 and G_2 are not isomorphic.

The inverse of the above corollary need not be true; see Problem 2-3.

<u>Def. 2-15.</u> The <u>complement</u> $\overline{G}$ of a graph G has $V(\overline{G}) = V(G)$ and $E(\overline{G}) = \{uv \mid u \neq v$ and $uv \notin E(G)\}$.

2-4. Operations on Graphs

We now define several binary operations on graphs. In what follows, we assume that $V(G_1) \cap V(G_2) = \emptyset$.

Def. 2-16.   1.) The <u>union</u> $G = G_1 \cup G_2$ has:
$V(G) = V(G_1) \cup V(G_2)$
$E(G) = E(G_1) \cup E(G_2)$.

Notation: $2G = G \cup G$
$nG = (n-1)G \cup G, \; n \geq 3$.

2.) The <u>join</u> $G = G_1 + G_2$ has:
$V(G) = V(G_1) \cup V(G_2)$
$E(G) = E(G_1) \cup E(G_2) \cup \{v_1 v_2 | v_i \in V(G_i), \; i = 1,2\}$.

3.) The <u>cartesian product</u> $G = G_1 \times G_2$ has:
$V(G) = V(G_1) \times V(G_2)$
$E(G) = \{[(u_1, u_2), (v_1, v_2)] | u_1 = v_1 \; \text{and} \; u_2 v_2 \in E(G_2)$
or $u_2 = v_2 \; \text{and} \; u_1 v_1 \in E(G_1)\}$.

4.) The <u>composition</u> (or <u>lexicographic product</u>)
$G = G_1[G_2]$ has:
$V(G) = V(G_1) \times V(G_2)$
$E(G) = \{[(u_1, u_2),(v_1, v_2)] | u_1 v_1 \in E(G_1)$ or
$u_1 = v_1 \; \text{and} \; u_2 v_2 \in E(G_2)\}$.

We are now in a position to conveniently define several infinite families of graphs.

Def. 2-17.   a.) P_n denotes the <u>path</u> of length n-1 (i.e. of order
n.)
b.) C_n denotes the <u>cycle</u> of length n.
c.) K_n denotes the <u>complete graph</u> on n vertices; that
is all $\binom{n}{2}$ possible edges are present.
d.) $\overline{K}_n$ denotes the <u>totally disconnected</u> graph on n
vertices; that is, $E(\overline{K}_n) = \emptyset$.

e.) $K_{m,n}$ denotes a <u>complete</u> <u>bipartite</u> graph:

$$K_{m,n} = \overline{K}_m + \overline{K}_n.$$

(Equivalently, $K_{m,n}$ is defined by:

$$\overline{K_{m,n}} = K_m \cup K_n.)$$

f.) $K_{p_1, p_2, \ldots, p_n}$ denotes a <u>complete</u> <u>n-partite</u> graph:

$$K_{p_1, p_2, \ldots, p_n} = \overline{K}_{p_1} + \overline{K}_{p_2} + \ldots + \overline{K}_{p_n},$$

an iterated join. In the special case where $p_1 = p_2 = \ldots = p_n$ ($= m$, say), we get a <u>regular</u> <u>complete</u> <u>n-partite</u> graph:

$$K_{m,m,\ldots,m} = K_n[\overline{K}_m].$$

g.) Q_n denotes the <u>n-cube</u> and is defined recursively:

$$Q_1 = K_2$$

$$Q_n = K_2 \times Q_{n-1}, \quad n \geq 2.$$

The complete bipartite graphs are a subclass of an extremely important class of graphs -- the bipartite graphs.

<u>Def. 2-18.</u> A <u>bipartite graph</u> G is a graph whose vertex set $V(G)$ can be partitioned into two non-empty subsets V' and V'' so that every edge of G has one vertex in V' and the other in V''.

<u>Thm. 2-19.</u> A graph G is bipartite if and only if all its cycles are even.

<u>Proof</u>: (i). Let $v_1 v_2 \ldots v_n v_1$ be a cycle in a bipartite graph G, and assume, without loss of generality, that $v_1 \in V'$; then $v_n \in V''$, and n must be even.

(ii). We may assume that G is connected, with only even cycles, since the argument in general follows directly from this special case. Consider a fixed v_0 $\in V(G)$. Let $V_i = \{u \in V(G) \mid d(u,v_0) = i\}$, $i = 0,1,$ $...,n$. Then n is finite, since G is connected, and $V_0, V_1, ..., V_n$ provides a partition of $V(G)$. Now, no two vertices in V_1 are adjacent, since G contains no 3-cycles. Also, no two vertices in V_2 are adjacent, or G would contain either a 3-cycle or a 5-cycle. In fact, every edge in G is of the form uv, where $u \in V_i$, $v \in V_{i+1}$, for some $i = 0,1,...,n-1$. Letting V' be the union of the V_i for i odd, and V'' be the union of the V_i for i even, we see that G is bipartite.

This completes our brief introduction; other terms will be defined, and theorems developed, as needed.

2-5. Problems

2-1.) Prove that if G is not connected, then $\overline{G}$ is connected. Give an example to show that the converse need not hold.

2-2.) A graph is said to be _perfect_ if no two vertices have the same degree. Prove that no graph is perfect, except $G = K_1$.

2-3.) Show that, even though $K_{3,3}$ and $K_2 \times K_3$ are both regular of order 6 and degree 3, they are not isomorphic.

2-4.) Prove that if G_1 and G_2 are both bipartite, then $G_1 \times G_2$ is also bipartite. Give an example to show that a similar result need not hold for the lexicographic product.

2-5.) Show that $\overline{G_1[G_2]} = \overline{G}_1[\overline{G}_2]$.

2-6.) Show that $\overline{G_1 + G_2} = \overline{G}_1 \cup \overline{G}_2$.

2-7.) Prove Theorem 2-9.

2-8.) Consider the graph G as a subspace of E^3. Show that G is connected as a topological space if and only if it is connected as a graph.

2-9.) Show that the first occurence of "connected" in Problem 2-8 may be replaced with "path-connected." (Recall that a path-connected topological space must be connected, but that the converse does not always hold. However, a connected and locally path-connected space must be path-connected.)

2-10.)   Prove Theorem 2-12.

2-11.)   Show that the set of all graphs, under the operation of
          cartesian product, forms a commutative semigroup with unity.

Chapter 3: The Automorphism Group of a Graph

In this chapter we show that there is associated, with each
graph, a group, known as the automorphism group of the graph. We
introduce various binary operations on permutation groups to aid in
computing automorphism groups of graphs. Several powerful results
relating graph and group products are stated, sometimes without
proof (see [27] for a further discussion); these results will not be
used in the sequel. Indeed, the concept of automorphism group of a
graph is, in the main, peripheral to the present course; it is introduced here primarily as one example of an interaction between
graphs and groups. (In Section 2 of Chapter 4 we find a more direct
bearing on subsequent material.)

3-1. Definitions

Def. 3-1. A one-to-one mapping from a finite set onto itself is
called a _permutation_. A _permutation group_ is a group
whose elements are all permutations acting on the same
finite set, called the _object set_. (The group operation
is composition of mappings.) If X is the object set
and A the permutation group, then $|A|$ is the _order_ of
the group, and $|X|$ is the _degree_.

Def. 3-2. Two permutation groups A and B are said to be _isomor-
phic_ $(A \cong B)$ if there exists a one-to-one onto map
$\theta: A \to B$ such that $\theta(a_1 a_2) = \theta(a_1)\theta(a_2)$, for all
$a_1, a_2 \in A$.

Def. 3-3. Two permutation groups A and B (acting on object sets
X and Y respectively) are said to be _identical_ $(A \equiv B)$
if: (i) $A \cong B$ (given by $\theta: A \to B$)

   (ii) there exists a one-to-one, onto map $f: X \to Y$
   such that $f(ax) = \theta(a)f(x)$, for all $x \in X$
   and $a \in A$.

Def. 3-4. An _automorphism_ of a graph G is an isomorphism of G
           with itself.   (The set of all automorphisms of G forms
           a permutation group, $G(G)$, acting on the object set
           $V(G)$.) $G(G)$ is called the _automorphism group_ of G.

Remark. An automorphism of G, which is a permutation of $V(G)$,
         also induces a permutation of $E(G)$, in the obvious
         manner.

Def. 3-5. An _identity graph_ is a graph G having trivial automor-
           phism group; that is, the identity permutation on $V(G)$
           is the only automorphism of G.

     It is easy to see that the graph pictured in Figure 3-1 is an
identity graph.   That there is no identity graph of smaller order
(other than K_1) is established in Problem 3-1.

G: 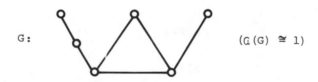 $(G(G) \cong 1)$

Figure 3-1

Thm. 3-6. $G(\overline{G}) \equiv G(G)$.

           Proof: Let $\theta: G(G) \rightarrow G(\overline{G})$ and $f: V(G) \rightarrow V(\overline{G})$ both be
           identity maps, and observe that adjacency is preserved in
           a graph if and only if non-adjacency is preserved.

           3-2. Operations on Permutation Groups

     From a theorem due to Cayley, we recall that any finite group
is abstractly isomorphic (as opposed to necessarily being identical)
with a permutation group; in fact, if the group G has order n,
then G is isomorphic to a subgroup of S_n.   In this light, the
operations soon to be defined could be regarded as applying to
groups in general; however, the definitions will be given in terms of
action upon a specified object set.

Let A and B be permutation groups acting on object sets X and Y respectively. We define three binary operations on these permutation groups as follows:

Def. 3-7. 1.) The <u>sum</u>, A + B, (or <u>direct product</u>) acts on the disjoint union $X \cup Y$; $A + B = \{a + b \mid a \in A, b \in B\}$, and

$$(a+b)(z) = \begin{cases} az, & \text{if } z \in X \\ \\ bz, & \text{if } z \in Y . \end{cases}$$

2.) The <u>product</u>, A × B, (or <u>cartesian product</u>) acts on $X \times Y$; $A \times B = \{a \times b \mid a \in A, b \in B\}$, and $(a \times b)(x,y) = ax, by$.

3.) The <u>composition</u>, A[B], (or <u>wreath product</u>) acts on X × Y as follows: for each $a \in A$ and any sequence $b_1, b_2, \ldots, b_d$ (where $d = |X|$) in B, there is a unique permutation in A[B], written $(a; b_1, b_2, \ldots, b_d)$, and $(a; b_1, b_2, \ldots, b_d)(x_i, y_j) = (ax_i, b_i y_j)$.

<u>Note</u>: The order of A[B] is $|A||B|^d$.

<u>Thm. 3-8</u>. $A + B \cong A \times B$.

<u>Proof</u>: Let $\theta: A + B \rightarrow A \times B$ be given by $\theta(a+b) = (a \times b)$.

Note that, in general, we cannot claim $A + B \equiv A \times B$.

3-3. Computing Automorphism Groups of Graphs

The following theorems indicate some connections between the graphical operations defined in Section 2-4 and the group operations defined above. The groups S_n, A_n, Z_n, D_n are respectively the symmetric and alternating groups of degree n, the cyclic group of order n, and the dihedral group of order 2n.

<u>Thm. 3-9</u>. If G is a connected graph, then $G(nG) \equiv S_n[G(G)]$.

Thm. 3-10. If no component of G_1 is isomorphic with a component
of G_2, then $G(G_1 \cup G_2) \equiv G(G_1) + G(G_2)$.

Proof: See Problem 3-2.

Thm. 3-11. Let $G = n_1 G_1 \cup n_2 G_2 \cup \ldots \cup n_r G_r$, where n_i is the
number of components of G isomorphic to G_i. Then

$$G(G) \equiv S_{n_1}[G(G_1)] + S_{n_2}[G(G_2)] + \ldots + S_{n_r}[G(G_r)].$$

Proof: Apply Theorems 3-9 and 3-10, using induction.

Note: Any graph G may be written as in Theorem 3-11, but if
$r = n_r = 1$, the theorem gives no information.

Thm. 3-12. If no component of $\overline{G}_1$ is isomorphic with a component
of $\overline{G}_2$, then $G(G_1 + G_2) \equiv G(G_1) + G(G_2)$.

Proof: Apply Theorems 3-6 and 3-10, together with
Problem 2-6.

The following two theorems are due to Sabidussi ([61] and [60]
respectively).

Def. 3-13. A non-trivial graph G is said to be prime if
$G = G_1 \times G_2$ implies that either G_1 or G_2 must be
trivial (i.e. $= K_1$). If G is not prime, G is
composite.

Thm. 3-14. $G(G_1 \times G_2) \equiv G(G_1) \times G(G_2)$ if and only if G_1 and G_2
are relatively prime.

Def. 3-15. The neighborhood of a vertex u is given by:
$N(u) = \{v \in V(G) | uv \in E(G)\}$. The closed neighborhood
is $N[u] = N(u) \cup \{u\}$.

Thm. 3-16. If G_1 is not totally disconnected, then $G(G_1[G_2]) \equiv$
$G(G_1)[G(G_2)]$, if and only if:

(i)   if there are two vertices in G_1 with the same
neighborhood, then G_2 is connected.

and

(ii)   if there are two vertices in G_1 with the same
closed neighborhood, then $\overline{G}_2$ is connected.

We are now able to list the automorphism groups for several
common families of graphs.

<u>Thm. 3-17</u>.   1.) $G(K_n) \equiv S_n$

2.) $G(C_n) \equiv D_n$

3.) $G(K_{m,n}) \equiv \begin{cases} S_2[S_n], & \text{if } m = n \\ S_m + S_n, & \text{if } m \neq n \end{cases}$

4.) $G(K_n[\overline{K}_m]) \equiv S_n[S_m]$.

3-4.   Graphs with a Given Automorphism Group

<u>Thm. 3-18</u>.   Every finite group is the automorphism group of some
graph.

For a proof of this theorem, due to Frucht [24], see
Section 4-2.

3-5.   Other Groups of a Graph

<u>Def. 3-19</u>.   Two graphs G and H (with non-empty edge sets) are
said to be <u>edge-isomorphic</u> if there exists a one-to-one,
onto map $\phi: E(G) \rightarrow E(H)$ preserving adjacency; that is
x,y share a vertex in G if and only if $\phi(x), \phi(y)$
share a vertex in H. ϕ is called an <u>edge-isomorphism</u>.

<u>Thm. 3-20</u>.   If G and H are isomorphic (with non-empty edge sets),
then they are edge-isomorphic.

<u>Proof</u>: See Problem 3-6.

<u>Def. 3-21</u>.   An <u>induced edge-isomorphism</u> is an edge isomorphism
$\phi: E(G) \rightarrow E(H)$ determined by $\phi(uv) = \theta u \theta v$,   where
$\theta: V(G) \rightarrow V(H)$ is an isomorphism.   An <u>edge-automorphism</u>
of a non-empty graph G is an edge-automorphism of G

with itself. (The set of all edge-automorphisms of G
forms a permutation group, $G_1(G)$ acting on the object
set $E(G)$.) $G_1(G)$ is called the <u>edge-automorphism</u>
<u>group</u> of G. Similarly, the <u>induced edge-automorphism</u>
<u>group</u> of G, $G^*(G)$, is the set of all induced edge-
automorphisms of G under composition.

We note that $G^*(G)$ is a (possibly proper; see Problem 3-8)
subgroup of $G_1(G)$. However, in almost every case, the three groups
introduced in this chapter are isomorphic (see, for example, [4]);
let $K_{1,3} + x$ denote the graph obtained by adding one edge to
$K_{1,3}$:

<u>Thm. 3-22.</u> For $G \neq K_1$, $G(G) \cong G^*(G)$ if and only if G contains
neither K_2 as a component nor two or more isolated
vertices.

<u>Thm. 3-23.</u> Let $E(G) \neq \phi$; then $G_1(G) \cong G^*(G)$ if and only if:

(1)   not both C_3 and $K_{1,3}$ are components of G, and

(2)   none of $K_{1,3} + x$, $K_4 - x$, K_4 is a component of G.

<u>Thm. 3-24.</u> Let G be a connected graph of order $p \geq 3$; then
$G(G) \cong G_1(G) \cong G^*(G)$ if and only if
$G \neq K_{1,3} + x$, $K_4 - x$, K_4 .

Thus it is customary to focus attention on the automorphism
group, $G(G)$.

3-6. Problems

3-1.) Does there exist an identity graph (other than K_1) of order
five or less? (Hint: Check to see that every graph in
Appendix 1, Harary [28], with $p \leq 5$, has at least one non-
trivial automorphism.)

3-2.) Prove Theorem 3-10.

3-3.) Prove Theorem 3-17.

3-4.) Let $G = K_{p,q,r}$; find $G(G)$.

3-5.) $G(K_4) \equiv S_4$, yet K_4 is the 1-skeleton of the tetrahedron, and the symmetry group of the tetrahedron is A_4. Explain!

3-6.) Prove Theorem 3-20.

3-7.) Show that the converse of Theorem 3-20 is false.

3-8.) Show that not all edge-automorphisms are induced.

Chapter 4: The <u>Cayley</u> <u>Color</u> <u>Graph</u> <u>of</u> <u>a</u> <u>Group</u> <u>Presentation</u>

In this chapter we see that any group may be defined in terms of generators and relations and that corresponding to such a presentation there is a unique graph, called the Cayley color graph of the presentation. A "drawing" of this graph gives a "picture" of the group, from which may be determined certain properties of the group. We will establish some basic results about Cayley color graphs, including a rather natural correspondence between direct products of groups and cartesian products of associated Cayley color graphs. In Chapter 7 we will ask which groups have Cayley color graphs that can be represented properly in the plane, and associated questions. In Chapter 9 appropriate answers will solve the Heawood map-coloring problem, as well as many others.

4-1. Definitions

<u>Def. 4-1</u>. Let Γ be a group, with $\{g_1, g_2, g_3, \ldots\}$ a subset of the element set of G. A <u>word</u> W in $g_1, g_2, g_3, \ldots$ is a finite product $f_1 f_2 \ldots f_n$, where each f_i is in the set $\{g_1, g_2, g_3, \ldots, g_1^{-1}, g_2^{-1}, g_3^{-1}, \ldots\}$. If every element of Γ can be expressed as a word in $g_1, g_2, g_3, \ldots$, then $g_1, g_2, g_3, \ldots$ are said to be <u>generators</u> for Γ. A <u>relation</u> is an equality between two words in $g_1, g_2, g_3, \ldots$.

<u>Thm. 4-2</u>. Given an arbitrary set of symbols and an arbitrarily prescribed set (possibly empty) of relations in these symbols, there is a unique (up to isomorphism) group with the symbols as generators and with structure determined by the prescribed relations.

(For a proof, see [40].)

Def. 4.3.   If Γ is generated by $g_1 , g_2 , g_3 , \ldots$ and if every
relation in G can be deduced from the relations P =
P', Q = Q', R = R', ..., then we write $\Gamma = \langle g_1 , g_2 ,$
$g_3 , \ldots |$ P = P', Q = Q', R = R', ... $\rangle$, and the right
hand side of the equation is said to be a presentation
of Γ. A presentation is said to be a finitely generated
(finitely related) if the number of generators (defining
relations) is finite. A finite presentation is both
finitely generated and finitely related.

Thm. 4-4.   Every finite group has a finite presentation.

Proof: Take Γ itself as the set of generators, with
all relations of the form $g_i g_j = g_k ,$ as determined by
the group operation. (i.e. the multiplication table
serves as a finite presentation.)

Def. 4-5.   For every group presentation there is associated a Cayley
color graph: the vertices correspond to the elements of
the group; next, imagine the generators of the group to
be associated with distinct colors. If vertices v_1 and
v_2 correspond to group elements g_1 and g_2 respectively, then there is a directed edge (of the color (or
label) of generator h) from v_1 to v_2 if and only if
$g_1 h = g_2 ;$ see Figure 4-1.

$$g_1 \overset{h}{\longrightarrow} g_1 h = g_2$$

Figure 4-1.

Let P be a presentation for the group Γ;   we denote the
Cayley color graph of P for Γ by $D_P(\Gamma)$,   or (when convenient)
by $D_\Delta(\Gamma)$, where Δ denotes the generating set. (Since a group
may have more than one generating set, the Cayley color graph de-
pends on Δ,   as well as Γ). Then $D_\Delta(\Gamma)$ is a labeled, directed graph, with a color (or label) assigned to each edge. We

observe that the following correspondences occur:

Group	Cayley Color Graph
element	vertex
generator	a set of directed edges of the same color
inverse of a generator	the same set of edges (now directed against the arrow)
word	walk
multiplication of elements	succession of walks
identity word	closed walk
solvability of $rx = s$	(weakly) connected di-graph

Note: A characterization is given in [19] of those graphs G which
        can be oriented and colored so as to form Cayley color graphs.

Historical note: Max Dehn (in 1911) formulated three fundamental de-
                  cision problems concerning group presentations.
                  One of these is: "determine in a finite number
                  of steps, for two arbitrary words W and W' in
                  the generators, whether W = W' or not." Equiv-
                  alently: "construct the Cayley color graph for a
                  given group presentation."

    The term "connected" may have a "stronger" meaning for directed
graphs than for graphs in general, since we may be allowed to travel
only in the direction of the arrow along a given directed edge.

Def. 4-6. A directed graph D is said to be strongly connected if,
           for every pair u, v of distinct vertices, there is a
           directed path from u to v. D is said to be unilat-
           erally connected if, for every pair of distinct vertices,
           one is joined to the other by a directed path. D is
           called weakly connected if the (undirected) pseudograph
           underlying D is connected.

    For example, see Figure 4-2, where D is strongly connected,
D' is unilaterally connected (but not strongly connected) and D"
is weakly connected (but not unilaterally connected).

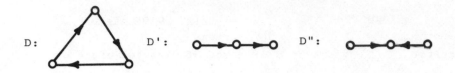

Figure 4-2.

4-2. Automorphisms

We have previously defined an automorphism of a graph G (as a permutation of $V(G)$ preserving adjacency). An automorphism of a directed graph must preserve <u>directed</u> adjacency; and an automorphism of a Cayley color graph must also preserve the <u>color</u> corresponding to each adjacency. We summarize in:

<u>Def. 4-7.</u>   An <u>automorphism of a Cayley color graph</u> $D_\Delta(\Gamma)$ is a permutation θ of $V(D_\Delta(\Gamma))$ such that, for each g_1, g_2 in Γ and h in Δ, $g_1 h = g_2$ if and only if $\theta(g_1)h = \theta(g_2)$.

Equivalently (see Problem 4-1), θ is an automorphism of $D_\Delta(\Gamma)$ if and only if: for each g in Γ and generator h in Δ, $\theta(gh) = \theta(g)h$; i.e. the diagram in Figure 4-3 commutes.

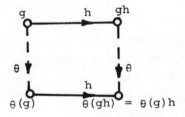

Figure 4-3

As expected, the collection of all automorphisms of $D_\Delta(\Gamma)$ forms a group, called the <u>automorphism group</u> of $D_\Delta(\Gamma)$, and denoted by $G(D_\Delta(\Gamma))$. The next result is perhaps not expected.

<u>Thm. 4-8.</u> Let $D_\Delta(\Gamma)$ be any Cayley color graph for the finite group
Γ; then $G(D_\Delta(\Gamma)) \cong \Gamma$ (independent of the presentation
selected for G.)

<u>Proof</u>: Define $\alpha: \Gamma \to G(D_\Delta(\Gamma))$ by $\alpha(g) = \theta_g$, where
$\theta_g: V(D_\Delta(\Gamma)) \to V(D_\Delta(\Gamma))$ is given by $\theta_g(g_i) = gg_i$.
First we show that $\theta_g \in G(D_\Delta(\Gamma))$. Clearly θ_g is one-
to-one and onto (and hence, permutes $V(D_\Delta(\Gamma))$). Also,
$\theta_g(g_ih) = g(g_ih) = (gg_i)h = \theta_g(g_i)h$, so that α is
well-defined.

Now, α preserves products; $\alpha(gg^*) = \theta_{gg^*}$, de-
fined by $\theta_{gg^*}(g_i) = gg^*g_i = \theta_g(g^*g_i) = \theta_g(\theta_{g^*}(g_i)) = (\theta_g\theta_{g^*})(g_i)$; that is $\alpha(gg^*) = \alpha(g)\alpha(g^*)$.

It is clear that α is one-to-one, since
ker $\alpha = \{e\}$.

It remains to show that α is onto. Let
$\theta \in G(D_\Delta(\Gamma))$. Let $\theta(e) = g$, where e is the identity
of Γ. Now any g* in Γ can be written as a word in
the generators for Γ; i.e. $g^* = h_1^{a_1}h_2^{a_2}\ldots h_m^{a_m}$, where
h_i is a generator for Γ and $a_i = \pm 1$. Then
$\theta(g^*) = \theta(eg^*) = \theta(e)h_1^{a_1}h_2^{a_2}\ldots h_m^{a_m} = gg^*$; that is,
$\theta = \theta_g$, so that α is onto. This completes the
proof.

From the above theorem (and its proof) it is evident that any
vertex of $D_\Delta(\Gamma)$ can be labeled with the identity e of Γ, and
that once this has been done (for fixed assignment of colors to the
generators) all other vertex labelings are determined.

We are now able to provide a proof of Frucht's Theorem, Theorem
3-18: Every finite group is the automorphism group of some graph.

<u>Proof</u>: Let Γ be a finite group, and let Δ be a generating
set for Γ. Form the Cayley color graph $D_\Delta(\Gamma)$; by Theorem 4-8,
we know that $G(D_\Delta(\Gamma)) \cong \Gamma$. It only remains to convert $D_\Delta(\Gamma)$
to a <u>graph</u> G having the same automorphism group, Γ. This is
done as follows: let $\Delta = \{\delta_1, \delta_2, \ldots, \delta_n\}$. Replace each edge
(g_i, g_j), where $g_j = g_i\delta_k$, by a path: $v_i, u_{ij}, u_{ij}', v_j$. At
vertex $u_{ij}(u_{ij}')$ we attach a new path P_{ij} (P_{ij}') of

length 2k-2 (2k-1); see Figure 4-4, for the case k = 2.   In
this way the "non-graphical" features of direction and label,
present in $D_\Delta(\Gamma)$, are incorporated into the graph G.   It is
clear that $G(G) \cong G(D_\Delta(\Gamma)) \cong \Gamma$.

Figure 4-4.

4-3. Properties

It is clear that every Cayley color graph is both regular and
connected (as a graph); the converse is not true (see Problem 4-10).
For two characterizations of graphs which may be regarded as Cayley
color graphs, see [32] and [40].

We may study additional properties for Γ (apart from the multiplication table so conveniently summarized in $D_\Delta(\Gamma)$) from $D_\Delta(\Gamma)$
as follows:

<u>Thm. 4-9</u>. Γ is commutative if and only if, for every pair of gene-
rators h_i and h_j,   the walk $h_i h_j h_i^{-1} h_j^{-1}$ is closed.

<u>Proof</u>: See Problem 4-2.

<u>Def. 4-10</u>. An element of a generating set for a group Γ is said
to be <u>redundant</u> if it can be written as a word in the
remaining generators. A generating set is said to be
<u>minimal</u> if it contains no redundant generators.

<u>Example</u>: $\{x^2, x^3\}$ is a minimal generating set for $Z_6 = \langle x | x^6 = e \rangle$,
even though $\{x\}$ is a generating set with fewer elements.

<u>Thm. 4-11</u>. A generator h is redundant if and only if the deletion
of all edges colored h in $D_\Delta(\Gamma)$ leaves a strongly
connected directed graph.

Proof: See Problem 4-3.

Thm. 4-12. If h is not redundant, the removal of all edges colored h leaves a collection of isomorphic disjoint
subgraphs, each representing the subgroup of Γ generated by the generating set of Γ minus h.

Proof: See Problem 4-4.

Thm. 4-13. Let Γ be a group with minimal generating set $\{h_1, h_2,$
$\ldots, h_r\}$, and Ω a (necessarily proper) subgroup with
generating set $\{h_2, h_3, \ldots, h_r\}$. Let $C_1, C_2, \ldots, C_k$
be the weak components of the directed graph $D_{h_1}(\Gamma)$,
obtained from $D_\Delta(\Gamma)$ by deleting the edges colored h_1.
Then Ω is normal in Γ if and only if the deleted
directed edges from any given component C_i all lead to
a single other component C_j.

Proof: (i) Assume the condition holds. Let $C_1 = \Omega$
be the component containing e, let $g \in C_1$ and
$r \in \Gamma$. We must show that $rgr^{-1} \in C_1$. We write $r =$
$a_1^{b_1} a_2^{b_2} \ldots a_m^{b_m}$, where a_i is a generator of Γ and
$b_i = \pm 1$. If h_1 occurs in r exactly w times with
$b_i = +1$ and v times with $b_i = -1$, then the walk
corresponding to r leads from e (in C_1) through
w-v components, ending in C_{1+w-v}. The walk corresponding to g now leads to another vertex in C_{1+w-v},
and the walk $a_m^{-b_m} \ldots a_2^{-b_2} a_1^{-b_1}$, corresponding to r^{-1},
returns us to C_1.

(ii) Suppose that edges colored h_1 lead from
C_i to C_1 and C_j, $1 \neq j$. (Again assume $e \in C_1$.)
Then there exists $g \in C_1$ such that $h_1^{-1} g h_1 \in C_j$, so
that Ω is not normal in Γ.

It now follows that, for Ω (as above) normal in Γ, the elements of the factor group Γ/Ω (i.e. the right cosets) are the
components of $D_{h_1}(G)$. By shrinking these components, each to a
single vertex, and restoring the edges colored h_1 (this can be
done unambiguously, by Theorem 4-13), a Cayley color graph of Γ/Ω

is <u>obtained</u>. (This "shrinking" may be described by adjoining, to the defining relations for G, the additional relations $h_2 = h_3 = \ldots = h_r = e$.)

In general, given a Cayley color graph $D_\Delta(\Gamma)$, whether a subgroup Ω of Γ is normal or not, we obtain a <u>Schreier</u> (right) <u>coset graph</u> as follows: the vertices are the right cosets of Ω in Γ, and there is an edge directed from Ωg to $\Omega g'$, labeled with $\delta \in \Delta$, if and only if $\Omega g \delta = \Omega g'$ (i.e. if and only if $\delta \in g^{-1}\Omega g'$.) That Ωgd is a right coset follows from the fact that the right cosets of Ω in Γ partition Γ. Note that a Schreier coset graph may actually be a pseudograph, as loops and/or multiple edges may result from this process. For the special case $\Omega = \{e\}$, the Schreier coset graph is just the Cayley graph $D_\Delta(\Gamma)$.

Several of the ideas discussed above are illustrated in Figure 4-5. Note that $\Omega = Z_3'$ is normal in S_3, but not in A_4.

As a further example, contrast the groups $Z_2 \times Z_4$ and D_4, as in Figure 4-6. Note that the subgroup of order 2 generated by r

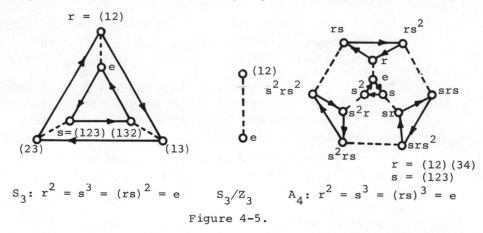

$$S_3: r^2 = s^3 = (rs)^2 = e \qquad S_3/Z_3 \qquad A_4: r^2 = s^3 = (rs)^3 = e$$

Figure 4-5.

is normal in $Z_2 \in Z_4$, but not in D_4. This comment extends in an obvious way to the groups $Z_2 \times Z_n$ and D_n, $n \geq 3$. For example, see $S_3 = D_3$ (in Figure 4-5); the subgroup generated by r is not normal here, either. For a generator δ of order 2, we adopt the standard convention of representing the two directed edges $(g,g\delta)$ and $(g\delta,g)$ in $D_\Delta(\Gamma)$ by a single undirected edge $[g,g\delta]$.

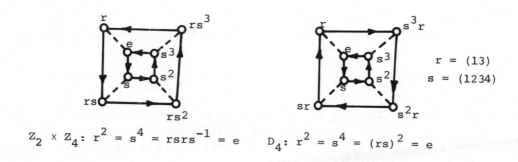

$$Z_2 \times Z_4: \quad r^2 = s^4 = rsrs^{-1} = e \qquad D_4: \quad r^2 = s^4 = (rs)^2 = e$$

$$r = (13)$$
$$s = (1234)$$

Figure 4-6

In Figure 4-7 we give the Schreier coset graph for $\Omega = \{e, r\}$, in $\Gamma = D_4$ (see Figure 4-6).

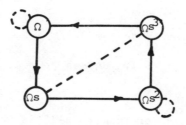

Figure 4-7.

4-4. Products

We now develop a relationship between the direct product for groups and the cartesian product for graphs. Recall the following from group theory:

<u>Def. 4-14.</u> Let Γ_1 and Γ_2 both be subgroups of the same group Γ, with $\Gamma_1 \cap \Gamma_2 = \{e\}$ and $gh = hg$ for all $g \in \Gamma_1$, $h \in \Gamma_2$. Then $\Gamma_1 \times \Gamma_2 = \{gh \mid g \in \Gamma_1, h \in \Gamma_2\}$ is also a a subgroup of Γ, called the <u>direct</u> <u>product</u> of Γ_1 and Γ_2.

If $\Gamma_1 = \langle k_1, \ldots, k_m \mid w_1 = \ldots = w_r = e \rangle$ and

$$\Gamma_2 = \langle k_{m+1}, \ldots, k_n \mid w_{r+1} = \ldots = w_{r+s} = e \rangle,$$

then $\Gamma_1 \times \Gamma_2 = \langle k_1, \ldots, k_n | \ w_1 = \ldots = w_{r+s} = k_i k_j k_i^{-1} k_j^{-1} = e,$

$$\text{for all} \quad 1 \leq i \leq m < j \leq n \rangle$$

is a presentation for $\Gamma_1 \times \Gamma_2$, called the _standard_ _presentation_
for $\Gamma_1 \times \Gamma_2$.

This binary operation may be extended to the class of all
groups, by noting that $\Gamma_1 \cong \Gamma_1' = \{(g, e_2) | \ g \in \Gamma_1, e_2$ is the iden-
tity of $\Gamma_2\}$, $\Gamma_2 \cong \Gamma_2' = \{(e_1, h) | \ h \in \Gamma_2, e_1$ is the identity of $\Gamma_1\}$,
and defining $\Gamma_1 \times \Gamma_2 = \{(g, h) | \ g \in \Gamma_1, h \in \Gamma_2\}$, with
$(g_1, h_1)(g_2, h_2) = (g_1 g_2, h_1 h_2)$ giving the group operation.

Also recall the following (see, for example [8, p. 348]):

<u>Thm. 4-15</u>.   (The Fundamental Theorem of Abelian Groups):
Let Γ be an abelian group of order n ; then $\Gamma =$
$Z_{m_1} \times Z_{m_2} \times \ldots \times Z_{m_r}$, where m_i divides m_{i-1}
$i = 2, \ldots, r$ and $\overset{r}{\underset{i=1}{\pi}} m_i = n$; furthermore, this
decomposition is unique.

(We assume $m_r > 1$, unless $n = 1$, in which case
$m_r = r = 1$.)

<u>Def. 4-16</u>.   The number r of Theorem 4-15 is called the <u>rank</u> of
the abelian group Γ .

Theorem 4-15 completely specifies the structure of finite abelian groups. The next theorem specifies, as a corollary, a Cayley
color graph for every finite abelian group. We first extend the definition of cartesian product for graphs to Cayley color graphs, in
the natural way.

<u>Def. 4-17</u>.   The <u>cartesian</u> <u>product</u>, $D_{\Delta_1}(\Gamma_1) \times D_{\Delta_2}(\Gamma_2)$, <u>of</u> <u>two</u>
<u>Cayley</u> <u>color</u> <u>graphs</u> is given by: $V(D_{\Delta_1}(\Gamma_1) \times D_{\Delta_2}(\Gamma_2)) =$
$V(D_{\Delta_1}(\Gamma_1)) \times V(D_{\Delta_2}(\Gamma_2))$; and (g_1, g_2) is joined to
(g_1', g_2') by an edge colored h if and only if either:

(i) $g_1 = g_1'$ and $g_2 h = g_2'$, for h a generator
in Δ_2

or

(ii) $g_2 = g_2'$ and $g_1 h = g_1'$,   for   h   a generator in Δ_1.

Figure 4-8 shows $D_{\Delta_1}(Z_3) \times D_{\Delta_2}(Z_2)$,   where $Z_3 = \langle x | x^3 = e \rangle$ and $Z_2 = \langle y | y^2 = e \rangle$.

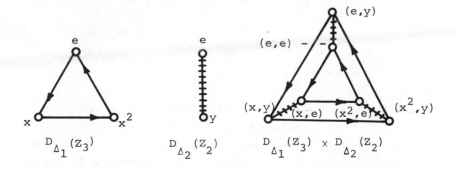

Figure 4-8.

<u>Thm. 4-18.</u>   Let $D_{P_i}(\Gamma_i)$ be the Cayley color graph associated with presentation P_i for group Γ_i, $i = 1, 2$. Let P be the standard presentation for $\Gamma_1 \times \Gamma_2$. Then

$$D_P(\Gamma_1 \times \Gamma_2) = D_{P_1}(\Gamma_1) \times D_{P_2}(\Gamma_2).$$

<u>Proof</u>: First we note that $V(D_{P_1}(\Gamma_1) \times D_{P_2}(\Gamma_2)) = V(D_{P_1}(\Gamma_1)) \times V(D_{P_2}(\Gamma_2)) = V(D_P(\Gamma_1 \times \Gamma_2))$. We now show that the edge sets of the two Cayley color graphs coincide (in <u>colored</u> <u>directed</u> <u>adjacency</u>.)

(i) Let (g_1, g_2) be joined to (g_1', g_2') by an edge colored h in $D_P(\Gamma_1 \times \Gamma_2)$. Then $h = k_i$, for some $1 \le i \le n$. If $1 \le i \le m$, then h is a generator of Γ_1, and

$$(g_1', g_2') = (g_1, g_2)(h, e_2) = (g_1 h, g_2),$$

so that $g_1' = g_1 h$ and $g_2' = g_2$; i.e. this directed, colored edge in $D_P(\Gamma_1 \times \Gamma_2)$ is also in $D_{P_1}(\Gamma_1) \times D_{P_2}(\Gamma_2)$. A similar argument applies for $m < i \le n$, so that

$$E(D_P(\Gamma_1 \times \Gamma_2)) \subseteq E(D_{P_1}(\Gamma_1) \times D_{P_2}(\Gamma_2)).$$

(ii)   The argument is reversible, to show that

$$E(D_{P_1}(\Gamma_1) \times D_{P_2}(\Gamma_2)) \subseteq D_P(\Gamma_1 \times \Gamma_2).$$

This completes the proof.

Since the cyclic group Z_n with presentation $P: Z_n = \langle x \mid x_n = e \rangle$, has the readily constructed Cayley color graph $D_P(Z_n) = C_n'$ (where C_n' denotes the directed cycle of length n), it is a simple matter to construct, using Theorems 4-15 and 4-18, a Cayley color graph for <u>any</u> finite abelian group.

<u>Thm. 4-19</u>.   Let Γ be a finite abelian group; then $C_{m_1}' \times C_{m_2}' \times \dots \times C_{m_r}'$ is a Cayley color graph for Γ, where $\Gamma = Z_{m_1} \times Z_{m_2} \times \dots \times Z_{m_r}$.

The class of groups for which we can construct Cayley color graphs -- using Theorem 4-18 -- can be enlarged as follows:

<u>Def. 4-20</u>.   A non-abelian group Γ is said to be <u>hamiltonian</u> if every subgroup of Γ is normal in Γ.

Clearly all abelian groups have this normality property for subgroups. That non-abelian groups may also have all subgroups normal is illustrated by Q, the quaternions (one of the two non-abelian groups of order eight). But more; the finite hamiltonian groups are characterized (see Coxeter and Moser [15], p. 8):

<u>Thm. 4-21</u>. Γ is a hamiltonian group if and only if $\Gamma = Q \times A_1 \times A_2$, where A_1 is an abelian group of odd order, and A_2 is a group for which $a^2 = e$, for every $a \in A_2$.

Since elementary group theory shows that A_2 must be abelian, we can apply Theorem 4-18 to find a Cayley color graph for Γ, providing we know a Cayley color graph for Q. This latter Cayley color graph will be produced in Chapter 7; it turns out to be the Cayley color graph of minimum order which cannot be drawn properly in the plane.

4-5. Problems

4-1.) Show that θ is an automorphism of $D_\Delta(\Gamma)$ if and only if: for each g in Γ and h in Δ, $\theta(gh) = \theta(g)h$.

4-2.) Prove Theorem 4-9.

4-3.) Prove Theorem 4-11.

4-4.) Prove Theorem 4-12.

4-5.) How many isomorphic disjoint subgraphs are there, as in the statement of Theorem 4-12?

4-6.) Give a graph-theoretic proof of the fact that a subgroup of index 2 must be normal.

4-7.) Let Γ be an abelian group of order pq, where p and q are distinct primes. By one of the Sylow theorems, Γ has Z_p as a subgroup. Give a graph-theoretic proof that Z_p is normal, first finding $D_P(Z_p \times Z_q)$. Then modify this Cayley color graph, to obtain a Cayley color graph of Γ/Z_p.

4-8.) Compile an appendix of Cayley color graphs, for all groups of order ≤ 12. (This appendix should be useful for reference , both in this course and in later life. You might want to save some work by doing Z_n, $Z_2 \times Z_n$, and D_n in general, rather than in each case where appropriate. Also. you will find that some of your graphs cannot be properly represented in the plane; these should be re-drawn, following Chapter 7.)

4-9.) Show that, if Γ is finite, then $D_\Delta(\Gamma)$ is always strongly connected. Give an example to show that this need not be true, if Γ is infinite.

4-10.) Show that the Petersen graph (see Figure 8-9) <u>cannot</u> be colored and labeled so as to be a Cayley color graph.

Chapter 5: <u>An Introduction to Surface Topology</u>

In this chapter we present an introduction to surface topology, including the statement and a brief discussion of the classification theorem for closed 2-manifolds and a complete development
of the euler formula for the orientable case. One motivation for
this material is that it gives us alternatives to the plane for
drawing graphs in (for example, no Cayley color graph for the
quaternions can be drawn properly in the plane); these alternatives
are completely classified, and the euler formula gives us important
information about them. We give a topological proof that there are
exactly five regular polyhedra, and conclude the chapter with a
brief discussion of pseudosurfaces.

5-1. Definitions

In this text, a <u>surface</u> will be a closed, orientable 2-
manifold. Any such figure may be considered as a topological subspace of euclidean 3-space, E^3. We consider the subspace topology
to be that induced by the standard distance-measuring metric in E^3.
To pin down this idea of "surface", we must define the terms used in
the first sentence of this paragraph. First, we specify that by the
<u>open</u> <u>unit</u> <u>disk</u> we mean $\overset{\circ}{D} = \{(x,y) \in E^2 \mid x^2 + y^2 < 1\}$.

<u>Def. 5-1</u>. A <u>2-manifold</u> is a connected topological space in which
every point has a neighborhood homeomorphic to the open
unit disk.

<u>Note</u>: In Definition 5-1, $\overset{\circ}{D}$ may be replaced by E^2, since these
two spaces are themselves homeomorphic.

<u>Example</u>: Only one of the conical spaces (the third) in Figure 5-1
is a 2-manifold.

Figure 5-1

Definition 5-1 may be extended as follows: an n-manifold is a connected topological space in which every point has a neighborhood homeomorphic to $B_n = \{(x_1, x_2, \ldots, x_n) \in E^n \mid \sum_{i=1}^{n} x_i^2 < 1\}$. However, we are only concerned here with the case $n = 2$.

Def. 5-2.   A subspace M of E^3 is bounded if there exists a natural number n such that $M \subseteq B(0;n) = \{(x,y,z) \mid x^2 + y^2 + z^2 < n\}$.

Def. 5-3.   Let $M \subseteq E^3$ be a 2-manifold. M is said to be closed if it is bounded and the boundary of M coincides with M.

For example, M of Figure 5-2 is closed, while M' and M'' are not.

$M = S^2$ $M' = S^2 - \bar{D}$ $M'' = E_2$

Figure 5-2

Note that the term "closed" does not mean quite the same thing to a surface topologist as it does to a point-set topologist. What a surface topologist calls a "closed 2-manifold", a point-set topologist calls a "compact 2-manifold." (Recall that $M \subseteq E^3$ is compact if and only if M is closed (point-set sense) and bounded.)

<u>Def. 5-4</u>. Let M be a 2-manifold; M is said to be <u>orientable</u> if,
         for every simple closed curve C on M, a clockwise
         sense of rotation is preserved by traveling once around
         C. Otherwise, M is <u>non-orientable</u>.

It can be shown that a 2-manifold M is orientable if and only
if it is two-sided. For example, a cylinder open at both ends is
orientable, whereas a Möbius strip is not.

5-2.   Surfaces and Other 2-manifolds

We finally know what a surface is (abstractly); now, exactly
which subspaces of E^3 are surfaces?
Let us begin to answer this question by representing certain
familiar 2-manifolds as polygons with appropriate edges identified.
See Figure 5-3 for the sphere, open cylinder, torus, projective
plane, möbius strip, and klein bottle, respectively. The top three
2-manifolds are orientable, the bottom three non-orientable. Only
the cylinder and möbius strip are not closed.

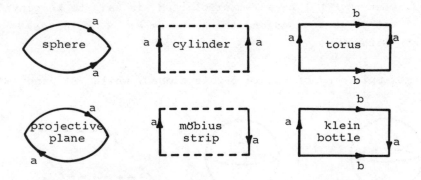

Figure 5-3

It turns out that every closed 2-manifold (whether orientable
or not) can be represented in this manner. In fact (see Fréchet
and Fan, [23] p. 63) we have the following theorem:

<u>Thm. 5-5</u>. Every closed 2-manifold is elementarily associated with
         a polygon whose symbolic representation is of one of the
         following forms:

(i) aa^{-1}

(ii) $a_1b_1a_1^{-1}b_1^{-1}a_2b_2a_2^{-1}b_2^{-1}\ldots a_pb_pa_p^{-1}b_p^{-1},p=1,2,3,\ldots$

(iii) $a_1a_1a_2a_2\ldots a_qa_q$, $q=1,2,3,$

The form (i) corresponds to the sphere;(ii) to the sphere with p handles (a torus is a sphere with one handle) and (iii) to the sphere with cross-caps (a projective plane is a sphere with one cross-cap; a klein bottle is a sphere with two cross-caps.) Only the forms (iii) correspond to non-orientable closed 2-manifolds.

As a byproduct of the development in Fréchet and Fan, an invariant called the characteristic is determined for each closed 2-manifold. Then it is shown that:

Thm. 5-6. (The Classification Theorem) Two closed 2-manifolds are homeomorphic if and only if they have the same characteristic and are both orientable or both non-orientable.

It follows that a closed orientable 2-manifold (i.e. a surface) M is a sphere with k handles, where k is a non-negative integer; k is said to be the genus of M, and we write $\gamma(M)=k$ and $M=S_k$.

5-3. The Characteristic of a Surface

We now give an independent determination of the characteristic of a surface, using the notion of a pseudograph. The proof will be by induction on k, the genus of the surface. We first need a few definitions, and one preliminary theorem. The first definition is intuitive; it will be made more precise in Chapter 6.

Def. 5-7. A pseudograph is said to be imbedded in a surface M if it is "drawn" in M so that edges intersect only at their common vertices.

For example, Figure 5-4 shows two drawings of K_4 in the plane, but only the second is an imbedding.

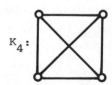 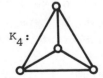

Figure 5-4

<u>Def. 5-8</u>.   A <u>tree</u> is a connected graph having no cycles.

For example, Figure 5-5 shows three graphs, but only G_3 is a
tree.

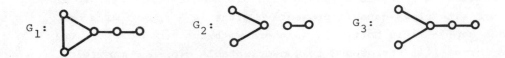

Figure 5-5

<u>Thm. 5-9</u>.   Let G be a tree, with p vertices and q edges; then
           p = q + 1.

           <u>Proof</u> (by induction on p) : The result is clearly true
           for p = 1, for then q = 0. Now assume the result holds
           for all trees with fewer than p vertices, and let G
           be a tree with p vertices and q edges. Since G has
           no cycles and is finite, we can find $v \in V(G)$ such that
           d(v) = 1. Then G - v is a tree with p - 1 vertices
           and q - 1 edges, so that (p-1) = (q-1) + 1; i.e.,
           p = q + 1.

<u>Def.5.10</u>.   Let a pseudograph G be imbedded in a surface M;   the
           components of M - G are called <u>regions</u> (or <u>faces</u>) of
           the imbedding.

For example, the imbedding of K_4 in Figure 5-4 has four
regions.

The following theorem is attributed to both Descartes and
Euler, independently; we perhaps indicate our preference by calling
it the <u>euler polyhedral formula</u>:

<u>Thm. 5-11</u>.   Let G be a connected graph imbedded in the sphere, S_0.
           Let G have p vertices and q edges, with r the
           number of regions of the imbedding. Then p - q + r = 2.

           <u>Proof</u> (by induction on q) : The result is clearly true
           for q = 1, for then p = 2 and r = 1. Now assume

the result holds for all connected graphs with fewer
than q edges, and let G be a connected graph with
q edges, p vertices, and r regions for an imbedding
in S_0. We have two cases to consider:

(i) If G is a tree, then $p = q + 1$ by Theorem
5-9, and $r = 1$ (since there are no cycles), so that
$p - q + r = 2$.

(ii) If G is not a tree, then (since G is connected) G contains a cycle; let x be any edge of
this cycle. Then G - x has p vertices, $q - 1$
edges, is still connected, and is imbedded in S_0 with
$r - 1$ regions. Hence $p - (q-1) + (r-1) = 2$; i.e.
$p - q + r = 2$.

Cor. 5-12. Let G be a connected pseudograph imbedded in S_0,
with p vertices, q edges, and r regions; then
$p - q + r = 2$.

Proof: See Problem 5-3.

We observe here that imbedding a graph in the sphere is equivalent to imbedding it in the plane. To see this, perform a stereo-
graphic projection (see Figure 5-6) with the north pole of the
sphere any point in the interior of some region of the imbedding.
For each point of the sphere, there corresponds a unique point of
the plane: the intersection of the line L through (0,0,2) and
(x,y,z) with the plane. The mapping is given explicitly by
$f: S^2 - P \rightarrow E^2$, where

$$S^2 = \{(x,y,z) \in E^3 \mid x^2 + y^2 + (z-1)^2 = 1\},$$

$$P = (0,0,2),$$

$$E^2 = \{(x,y,z) \in E^3 \mid z = 0\},$$

and

$$f(x,y,z) = (x',y',0),$$

with

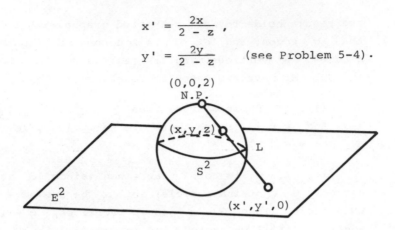

$$x' = \frac{2x}{2 - z} ,$$

$$y' = \frac{2y}{2 - z} \qquad \text{(see Problem 5-4)}.$$

Figure 5-6

The image of the graph G from S^2 is an imbedding of G in E^2, with the unbounded region corresponding to the region in S^2 from the interior of which the north pole was selected. Clearly, this process is reversible. In fact, the map f gives a homeomorphism between $S^2 - P$ and E^2, where P is any point of S^2. Note that neither space is closed from the point of view of surface topology, yet E^2 (and _not_ $S^2 - P$) is closed -- in the point-set sense -- as a subspace of E^3.

Def. 5-13.   A region of an imbedding of a graph G in a surface M is said to be a <u>2-cell</u> if it is homeomorphic to the open unit disk. If every region for an imbedding is a 2-cell, the imbedding is said to be a <u>2-cell imbedding</u>.

Thm. 5-14.   Let G be a connected pseudograph, with a 2-cell imbedding in S_k, with the usual parameters p, q, and r. Then

$$p - q + r = 2 - 2k.$$

Proof (by induction on k); The case k = 0 has been settled by Corollary 5-12. Now assume the theorem is true for fewer than k handles (k ≥ 1), and let G be as in the statement of the theorem. Without loss of generality, we assume all the vertices of G to be on the "sphere" portion of S_k; and since the imbedding is 2-cell, each handle has at least one edge of G running over it. Select one handle, and draw two dis-

joint simple closed curves C_1 and C_2 around this
handle. Suppose edges $x_1, x_2, \ldots, x_n$ run over the
handle, where $n \geq 1$. Then C_i meets x_j in a point
of S_k which we designate by u_{ij}, $i = 1,2$;
$j = 1,2,\ldots,n$. Consider the points u_{ij} to be vertices
of a new pseudograph, with edges determined in the
natural manner. Now remove the portion of the handle
between C_1 and C_2 and "fill in" the two resulting
holes (bounded by C_1 and C_2 respectively) with two
disks (this is called a <u>capping</u> operation). The result
is a 2-cell imbedding of a connected pseudograph in
S_{k-1}, with parameters p', q', and r' (say). But

$$p' = p + 2n$$

$$q' = q + 3n$$

$$r' = r + n + 2.$$

Thus, by the inductive assumption,

$$2 - 2(k-1) = p' - q' + r'$$

$$= (p+2n) - (q+3n) + (r+n+2)$$

$$= p - q + r + 2;$$

that is, $p - q + r = 2 - 2k$.

<u>Cor. 5-15</u>.   Let G be a connected graph, with a 2-cell imbedding
in S_k, with the parameters p, q, and r; then
$p - q + r = 2 - 2k$.

<u>Proof</u>: The result is immediate, since any graph is also
a pseudograph.

We have shown that the number $p - q + r$ is invariant for
S_k, for any 2-cell imbedding of any connected pseudograph;
$p - q + r = 2 - 2k$, depending only on k. This invariant number,
$2 - 2k$, is called the <u>euler</u> <u>characteristic</u> for the surface S_k.
It then follows that S_n and S_m are homeomorphic if and only if
$m = n$. In the non-orientable case, the characteristic is given by
$p - q + r = 2 - k$, where k is the number of cross-caps (see
Fréchet and Fan.)

5-4. Two Applications

The ramifications of Theorem 5-14 are enormous. In the remainder of this section, we give only two of these, both pertaining to the case $k = 0$.

Def. 5-16. A graph is said to be <u>planar</u> if it can be imbedded in the plane (or, equivalently, in S_0). A graph imbedded in S_0 is called a <u>plane</u> graph.

Notation: Suppose a graph G is 2-cell imbedded in a surface S_k . Let v_i be the number of vertices of degree i, and let r_i designate the number of regions having i sides (i.e. the number of regions having as boundary a closed walk of length i). We assume that $v_0 = v_1 = v_2 = 0$, as we focus on polyhedral graphs in this section; moreover, since G is a graph, $r_0 = r_1 = r_2 = 0$.

Lemma 5-17. (i) $p = \sum_{i \geq 3} v_i$

(ii) $r = \sum_{i \geq 3} r_i$

(iii) $2q = \sum_{i \geq 3} iv_i$

(iv) $2q = \sum_{i \geq 3} ir_i$

Proofs: (i) and (ii) are obvious; (iii) is Theorem 2-2, and (iv) follows in like manner to (iii); in summing the number of sides in the regions, each edge is counted exactly twice.

Thm. 5-18. The graph K_5 is not planar.

Proof: Suppose that the connected graph K_5 were imbedded in the plane; then $2q = 20 = \sum_{i \geq 3} ir_i \geq 3 \sum_{i \geq 3} r_i$ = 3r,   and by Theorem 5-11,

$$2 = p - q + r$$

$$\leq 5 - 10 + \frac{20}{3} = \frac{5}{3},$$

a contradiction! Hence, K_5 is not planar.

<u>Lemma 5-19</u>. Let the planar connected graph G be imbedded in the
plane; then

> (i)   G has a vertex of degree 5 or less; <u>and</u>

> (ii) G has a region with 5 or less sides.

> <u>Proof</u>: (i) Suppose, to the contrary, that $v_i = 0$,
> i 0,1,2,3,4 5; then $2q = \sum\limits_{i\geq3} iv_i \geq 6 \sum\limits_{i\geq3} v_i = 6p$. As
> before, $2q = \sum\limits_{i\geq3} ir_i \geq 3 \sum\limits_{i\geq3} r_i = 3r$. Then, by Theorem
> 5-11, $p - q + r = 2$; i.e.

> $q = p + r - 2$

> $\leq q/3 + 2q/3 - 2 = q - 2$,

a contradiction.

> (ii) This follows by duality (soon to be
> explained); it also follows from Problem 5-6.

We are now prepared to give a <u>topological</u> proof of what the
Greeks knew, geometrically, over two thousand years ago: there are
exactly five regular polyhedra. A <u>polyhedron</u> is a finite, connected
collection of at least three polygons, fit together in E^3 so that:
(i) each side of each polygon coincides exactly with one side of one
other polygon, and (ii) around each vertex there is one circuit of
polygons; together with the region of E^3 bounded by these polygons. These two conditions rule out the anomalies depicted in
Figure 5-7.

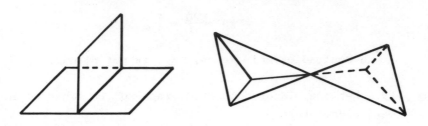

Figure 5-7

A <u>regular</u> <u>polyhedron</u> is a convex polyhedron for which: (i) the polygons are congruent regular polygons, and (ii) the same number of
polygons surround each vertex.

<u>Thm. 5-20</u>. There are exactly five regular polyhedra.

> <u>Proof</u>: Let P be a regular polyhedron. Associated
> with P is a regular planar graph G (to picture this,
> first bound P with a sphere, then place a light source
> inside the polyhedron -- the shadow of the vertices and
> edges of P gives a graph imbedded in the sphere;
> finally, perform a stereographic projection.) This pla-
> nar graph G has $v_0 = v_1 = v_2 = 0$, and in fact:
> $p = v_k$, $r = r_h$, for $k,h \in \{3,4,5\}$, by Lemma 5-19.
> Next, by Theorem 5-11, $p - q + r = 2$; we re-write this
> as follows:
>
> $$8 = 4p + 4r - 2q - 2q$$
>
> $$= \sum_{i \geq 3} (4-i)(r_i + v_i)$$
>
> $$= (4-h)r_h + (4-k)v_k.$$
>
> But also, $hr_h = kv_k$, since both $= 2q$, by Lemma 5-17.
> Of the nine possibilities for (h,k) in positive
> integers, only the following satisfy both of the above
> equations in r_h and v_k: $(h,k) =$
>
>   (i) $(3,3)$; $r_3 = v_3 = 4$ (the <u>tetrahedron</u>)

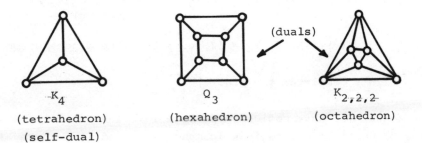

K_4
(tetrahedron)
(self-dual)

Q_3
(hexahedron)

(duals)

$K_{2,2,2}$
(octahedron)

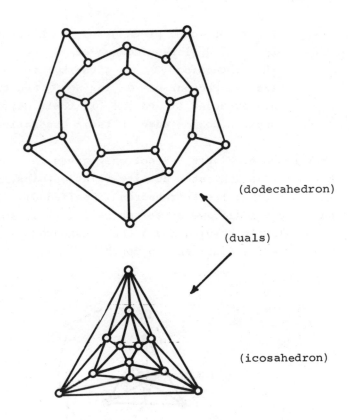

(dodecahedron)

(duals)

(icosahedron)

Figure 5-8

(ii)   (3,4); $r_3 = 8$, $v_4 = 6$ (the <u>octahedron</u>)

(iii)   (3,5); $r_3 = 20$, $v_5 = 12$ (the <u>icosahedron</u>)

(iv)   (4,3); $r_4 = 6$, $v_3 = 8$ (the <u>hexahedron</u>; i.e.
                                          the cube)

(v)   (5,3); $r_5 = 12$, $v_3 = 20$ (the <u>dodecahedron</u>)

This completes the proof.

The reader may have noticed a certain interchangability between
the roles of vertices and regions (compare (ii) and (iv) above, (iii)
and (v) above; see Lemmas 5-17 and 5-19, and Theorem 5-14). This is
no accident.

<u>Def. 5-21.</u> Let a connected pseudograph G be 2-cell imbedded in
          S_k. The <u>dual</u> pseudograph of G, $D_I(G)$ (relative to
          this imbedding I), is given by: the vertices of $D_I(G)$
          are the regions of G in S_k, and two such vertices
          are adjacent if and only if their corresponding regions
          share a common edge in their boundaries.

For example, Figure 5-8 not only gives the regular polyhedral
graphs, but also indicates duality relationships. Figure 5-9 shows,
for instance, that the tetrahedron is self-dual. Thus, having established Lemma 5-9 (i), we establish part (ii) by applying (i) to the
dual. Similarily, having found the hexahedron, we discover the
octahedron as its dual; and so forth.

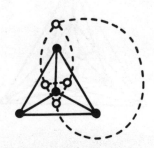

Figure 5-9

Although there are only five regular polyhedra (also called the
Platonic Solids), there are infinitely many convex polyhedra, as the
classes of all prisms and antiprisms show. The thirteen Archimedean
Solids are also all convex polyhedra. There are many non-convex
polyhedra, some of which are uniform with regard to face structure
and vertices, which have planar graphs as 1-skeletons; see, for
example, [69]. For a splendid study of orientable polyhedra (of
possibly positive genus) with regular faces, consult B. M. Stewart's
Adventures Among the Toroids [64].

Def. 5-22. A graph G is said to be 3-polytopal if it is the
            1-skeleton of a convex polyhedron.

Def. 5-23. A graph G is said to be n-connected (n ≥ 1) if the
            removal of fewer than n vertices from G neither
            disconnects G nor reduces G to the trivial graph
            K_1.

Graphs which are 3-polytopal have been characterized by
Steinitz [63].

Thm 5-24. A graph G is 3-polytopal if and only if it is planar
           and 3-connected.

One readily verifies that the five planar graphs of Figure 5-8
are also 3-connected. The following theorem of Weinberg [68] gives
information about the automorphism groups of 3-polytopal graphs:

Thm 5-25. Let G be 3-polytopal, with q edges. Then $|\mathcal{A}(G)| \le 4q$,
           with equality holding if and only if G is the 1-
           skeleton of a Platonic Solid.

In the next chapter we consider imbedding graphs (and graphs
of groups!) in surfaces of positive genus.

5-5. Pseudosurfaces

We now consider topological spaces akin to surfaces, but which fail to be 2-manifolds at a finite number of points; these spaces form additional candidates for the imbedding of graphs, and were studied extensively by Petroelje [49].

Def.5-26. Let A denote a set of $\sum_{i=1}^{t} n_i m_i \geq 0$ distinct points of S_k, with $1 < m_1 < m_2 < \ldots < m_t$. Partition A into n_i sets of m_i points each, $i = 1, 2, \ldots, t$. For each set of the partition, identify all the points of that set. The resulting topological space is called a pseudo-surface, and is designated by $S(k; n_1(m_1), n_2(m_2), \ldots, n_t(m_t))$. Each point resulting from an identification of m_i points of S_k is called a singular point. If a graph G is imbedded in a pseudosurface, we assume that each singular point is occupied by a vertex of G; such a vertex is called a singular vertex.

Thm. 5-27. Let G be a graph having a 2-cell imbedding in $S(k; n_1(m_1), n_2(m_2), \ldots, n_t(m_t))$; then $p - q + r = 2 - 2k - \sum_{i=1}^{t} n_i(m_i - 1)$.

The number $2 - 2k - \sum_{i=1}^{t} n_i(m_i - 1)$ is said to be the character-istic for the pseudosurface, and is a topological invariant, just as $2 - 2k$ is for the surface S_k.

5-6. Problems

5-1.) A forest is a graph for which every component is a tree. Show that, if G is a forest with p vertices, q edges, and k components, then $p = q + f(k)$, where $f(k)$ must be determined.

5-2.) Let G, a graph with p vertices, q edges, and k components, be imbedded in the sphere, with r regions. Show that $p - q + r = g(k)$, where $g(k)$ must be determined. Illustrate, for $G = 2K_4$. For what values of k will the imbedding be 2-cell?

5-3.)   Prove Corollary 5-12.

5-4.)   Verify that $f(x,y,z) = (2x/(2-z), 2y/(2-z), 0)$ gives the
         stereographic projection.

5-5.)   Show that $K_{3,3}$ is not planar.

5-6.)   Prove Lemma 5-19 (ii), without using duality.

5-7.)   Where might the proof of Theorem 5-14 break down, for graphs
         (instead of pseudographs)?

5-8.)   Consider the 2-manifolds in Figure 5-3. Determine in which
         of these K_5 can be imbedded. For each 2-manifold, compute
         the characteristic. Note that the characteristics agree for
         the torus and the klein bottle; are these two homeomorphic?
         Why? The characteristics also agree for the möbius strip
         and the projective plane; are they homeomorphic? How about
         the sphere and the cylinder?

5-9.)   Show that the two symbolic representations $ab^{-1}ab$ (as in
         Figure 5.3) and $a_1a_1a_2a_2$ (as in Figure 5-5 (iii)) both
         give the klein bottle. (Hint: cut along an appropriate
         diagonal of the rectangle $a_1a_1a_2a_2$ and then make an
         appropriate identification to obtain the rectangle $ab^{-1}ab$.)

5-10.)   For G the 1-skeleton of a Platonic Solid, show that
         $|G(G)| = 4q$.

5-11.)   The <u>wheel</u> graph W_m is defined as the join (see Def. 2-10)
         $K_1 + C_{m-1}$, $m \geq 4$. Show that $|G(W_m)| \leq 4q$, with equality
         holding iff m = 4. Is this consistent with Thm. 5-25?

*
5-12.)   Give an example to show that two pseudosurfaces with the
         same characteristic can be non-homeomorphic (compare the
         situation for surfaces). Find a formula that gives, for
         $n \geq -2$, the number of non-homeomorphic pseudosurfaces with
         characteristic -n.

Chapter 6: Imbedding Problems in Graph Theory

Recall from Definition 2-1 that a graph is an abstract mathematical system. It is when we concern ourselves with the geometric
realization of a graph as a finite one-dimensional complex that imbedding problems arise. There are practical applications for this
view of graphs. For instance, we will see in Chapter 8 that one of
the truly famous problems in mathematics can be stated in terms of
imbedded graphs. As another example, imagine the task of printing
an electronic circuit on a circuit board. Associated with the
circuit (in an obvious manner) is a graph, and the circuit can be
printed without shorts if and only if the associated graph can be
imbedded in the plane. What to do if the graph is not planar will
be studied in this chapter.

What do we mean by "the geometric realization" of a graph?   In
this section, we will normally mean a configuration in E^3, where
the vertices of the graph are represented by distinct points, and
the edges of the graph by lines; two lines intersect only at a point
representing common end vertices of the two corresponding edges. A
natural question is: "in what subspaces of E^3 will a given graph
imbed in this manner?" We will confine our attention to the following subspaces:

(i) E^3 itself
(ii) E^2
(iii) n-books (see definition below)
(iv) surfaces
(v)   pseudosurfaces.

Def. 6-1. An n-book is the cartesian product of the unit interval
with the geometric realization of the graph $K_{1,n}$.

That is, an n-book consists of n rectangles (the pages)
joined along a common edge (the spine).

6-1.   Answers to Some Imbedding Questions

The imbedding question has been completely answered for (i),
(ii), and (iii), as the next three theorems indicate. We will also need
some definitions. In the sequel, the term "graph" will be used in-
terchangeably, to represent either the abstract mathematical system,
or a realization of this system in E^3. The context should make it
clear which use is intended.

Thm. 6-2.   If K is a countable and locally finite simplicial com-
            plex, with dim K $\leq$ n,   then K has a realization (i.e.
            a linear imbedding) as a closed subset in E^{2n+1}.

            (See Spanier [62] for a discussion of this theorem.)

Cor. 6-3.   Any finite one-complex is imbeddable in E^3.

Note that Corollary 6-3 indicates that any graph may be imbedded in E^3,   and in such a way that every edge is represented as a
straight line.

Another way to see this is as follows. Let C be the curve in
E^3 determined by the parametric equations $x = t$, $y = t^2$, $z = t^3$
($t \geq 0$). Select p distinct points along C to represent the
vertices of G and represent the q edges of G as straight lines
joining these points appropriately. Since no four points on C are
coplanar, C meets any plane in E^3 at most three times, and no
two edges of G intersect extraneously.

Def. 6-4.   An elementary subdivision of an edge uv of a graph is
            the deletion of edge uv, the addition of a new vertex
            w, and the addition of two new edges, uw and wv.

Def. 6-5.   A graph G is said to be homeomorphic from a graph H
            if G can be obtained from H by a (finite) sequence of
            elementary subdivisions. (We say that G is a subdivi-
            sion of H.) G_1 and G_2 are said to be homeomorphic
            with each other if they are both homeomorphic from a
            common graph H.

Note that G_1 is homeomorphic with G_2 in the graph-theoretical sense defined above if and only if the realizations of G_1 and

G_2 in E^3 are homeomorphic in the topological sense (see Problem 6-1.)

The next theorem is one of the most important in all of graph theory; it is due to Kuratowski [36].

Thm. 6-6. A graph G is planar if and only if it contains no sub-
        graph homeomorphic with either K_5 or $K_{3,3}$.

It is clear that any graph with q edges can be imbedded in a q-book: place all the vertices along the spine, and use one page for each edge. However, we can do much better; the following theorem, due to Atneosen [2], is rather suprising.

Thm. 6-7. Any graph G can be imbedded in a 3-book.

Note that by Theorem 6-6, we have a criterion for ascertaining if the third page is needed for a particular graph. Theorem 6-7 is proved as follows: as shown in Massey [42], any closed 2-manifold with non-void boundary can be represented as a disk with strips attached in a certain way. Clearly any graph G can be imbedded in a closed 2-manifold with non-void boundary (simply remove an open disk from the interior of some region, for any S_k in which G can be imbedded; take $k = q$, for example). Atneosen showed, very neatly, that any disk with strips attached as described by Massey can be imbedded in a 3-book.
        So, it is only for the subspaces (iv) and (v) of our list above --that is, the surfaces and pseudosurfaces -- that the imbedding problem is, in general, unsolved, for non-planar graphs. Clearly, any graph will imbed on S_k, for k large enough (for example, take $k = q$ and use one handle for each edge); but this does not characterize which graphs imbed on S_k, for k fixed. The most natural problem here might be: for a given graph, find the surface of minimum genus in which the graph can be imbedded. If the graph is associated with an electronic circuit, the corresponding problem is: find the fewest number of holes that must be punched in the circuit board so that the board can accommodate the circuit. For Cayley color graphs, the problem becomes: find the simplest locally 2-dimensional "drawing board" in which to "paint" a picture of a given group. We will also see, in Chapter 8, that imbedding certain graphs in appropriate surfaces will tell us a good deal about map-coloring problems.

6-2. Definition of "Imbedding"

Let us now give two very careful definitions of "imbedding"
(they are easily seen to be equivalent), and then proceed to study
this process in some detail.

<u>Def. 6-8</u>. Let G be a graph, with $V(G) = \{v_1, v_2, \ldots, v_n\}$ and
$E(G) = \{x_1, x_2, \ldots, x_m\}$. Let M be a 2-manifold. An
<u>imbedding</u> <u>of</u> G <u>in</u> M is a subspace G(M) of M such
that

$$G(M) = \bigcup_{i=1}^{n} v_i(M) \cup \bigcup_{j=1}^{m} x_j(M),$$

where
  (i) $v_1(M), \ldots, v_n(M)$ are distinct points of M
  (ii) $x_1(M), \ldots, x_m(M)$ are m mutually disjoint open
         arcs in M
  (iii) $x_j(M) \cap v_i(M) = \phi$, i = 1,...,n; j = 1,...,m.
  (iv)   if $x_j = (v_{j1}, v_{j2})$, then the open arc $x_j(M)$
         has $v_{j1}(M)$ and $v_{j2}(M)$ as end points;
         j = 1,...,m.

In the above definition, an <u>arc</u> in M is a homeomorphic image
of [0,1]; an open arc is an arc less its two end points, the
images of 0 and 1.
    Equivalently (and much more briefly) we have:

<u>Def. 6-8'</u>. The graph G can be <u>imbedded</u> in the 2-manifold
         M if the geometric realization of G as a one-dimen-
         sional simplicial complex is homeomorphic to a subspace
         of M.

6-3. The Genus of a Graph

Imbedding question (iv) in this chapter leads directly to:

<u>Def. 6-9</u>. The <u>genus</u>, $\gamma(G)$, <u>of a graph</u> G is the minimum genus
         among all surfaces in which G can be imbedded.

For example, if G is planar then we write $\gamma(G) = 0$. If
$\gamma(G) = k$, $k > 0$, then G has an imbedding in S_k, but not in
S_h, for $h < k$. Moreover, G imbeds in S_m, for all $m \geq k$
(merely add $m - k$ handles to an imbedding of G in S_k).

As mentioned above, it is clear that every graph has a genus.
Let G have q edges; then place the vertices of G on the sphere,
and add one handle for each edge. Thus $\gamma(G) \leq q$.

<u>Def. 6-10</u>. An imbedding of a graph G in a surface S_k is said to
be <u>a</u> <u>minimal</u> <u>imbedding</u> if $\gamma(G) = k$.

The next result is extremely useful, as it tells us that the
euler formula applies for any minimal imbedding of a connected
graph. For a complete proof, see [77].

<u>Thm. 6-11</u>. If a connected graph G is minimally imbedded in a
surface, then the imbedding is a 2-cell imbedding.

<u>Heuristic Argument</u>: We assume (without loss of gener-
ality) that every vertex of G lies on the sphere.
Hence only edges can be imbedded on the handles. Suppose that R is a non- 2-cell region. Then there is a
simple closed curve C in R which cannot be continu-
ously deformed, in R, to a point. If $\gamma(G) = 0$, C
divides S_0 into two parts (by the Jordan curve
theorem), each of which must contain a vertex of G.
But then G would be disconnected. Hence $\gamma(G) \geq 1$.
We consider three cases:

<u>Case (i)</u>. If C lies entirely on one handle, we cut
the surface along C, cap the two resulting holes, and
obtain an imbedding of G in $S_{\gamma(G)-1}$, a contradiction.

<u>Case (ii)</u>. If C lies entirely on the sphere, we re-
gard the "sphere" portion of the surface as a handle,
and apply case (i).

<u>Case (iii)</u>. If C lies partially on some handle H
and partially on $S_{\gamma(G)}$ - H, we redraw the edges of G
formerly carried by H along that portion of C lying
in $S_{\gamma(G)}$ - H; we obtain an imbedding of G on the sur-

face without using handle H, the final contradiction.

The corollary below follows directly from Theorems 5-14 and
6-11.

Cor. 6-12. If a connected graph G has a minimal imbedding in S_k ,
with p vertices, q edges, and r regions, then

$$p - q + r = 2 - 2k.$$

The next two corollaries are often helpful in computing the
genus of a graph. We require two new terms.

Def. 6-13. A 2-cell imbedding is said to be a triangular (quadri-
lateral) imbedding if $r = r_3$ $(r = r_4)$.

Cor. 6-14. If G is connected, with p ≥ 3, then $\gamma(G) \geq \frac{q}{6} - \frac{p}{2} + 1$.
Furthermore, equality holds if and only if a triangular
imbedding can be found for G.

Proof: Let G be imbedded in $S_{\gamma(G)}$, so that
$p - q + r = 2 - 2\gamma(G)$. Since $2q \geq 3r$, with equality
if and only if $r = r_3$ (see Lemma 5-17), the result is
immediate.

Cor. 6-15. If G is connected, with p ≥ 3, and has no triangles,
then $\gamma(G) \geq \frac{q}{4} - \frac{p}{2} + 1$. Furthermore, equality holds if
and only if a quadrilateral imbedding can be found for
G.

(The proof is entirely analogous to that of Cor. 6-14.)

The next corollary will be heavily used in the remainder of
this chapter.

Cor. 6-16. If G is a connected bipartite graph having a quadrilateral imbedding, then $\gamma(G) = \frac{q}{4} - \frac{p}{2} + 1$.

Proof: Apply Theorem 2-19 and Corollary 6-15.

We have shown (among other things) that, for connected graphs,
minimal imbeddings are 2-cell imbeddings. Two questions arise:

(i) what about minimal imbeddings of disconnected graphs?   (ii) Are
there 2-cell imbeddings which are not minimal? We discuss these
two questions briefly.

Def. 6-17.   Given a connected graph G,   a cut-vertex is a vertex v
              such that G - v is disconnected. A block is a maximal
              connected subgraph of G having no cutvertices.

    For example, the graph in Figure 6-1 has two blocks, both iso-
morphic to K_4;   v is a cutvertex for this graph.   Note that a
block is either K_2 or is 2-connected.

G:
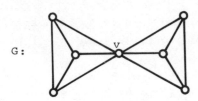

Figure 6-1.

    The next theorem and its corollary are due to Battle, Harary,
Kodama, and Youngs [3], and are presented without proof.

Thm. 6-18.   The genus of a graph is the sum of the genera of its
              blocks.

Cor. 6-19.   The genus of a graph is the sum of the genera of its
              components.   (i.e., let $G = \bigcup\limits_{i=1}^{n} C_i$;   then $\gamma(G) =$
              $\sum\limits_{i=1}^{n} \gamma(C_i)$).

6-4.   The Maximum Genus of a Graph

    That there exist 2-cell imbeddings which are not minimal is
evident from Figure 6-2, which shows K_4 in S_1.   Note that the
euler formula still applies here (4 - 6 + 2 = 0).   It is clear
that no imbedding of a disconnected graph can be a 2-cell imbedding.
To describe all 2-cell imbeddings of a given connected graph, we

introduce the following concept:

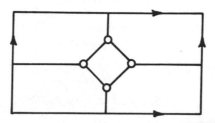

Figure 6-2

Def. 6-20. The <u>maximum genus</u>, $\gamma_M(G)$, <u>of a connected graph</u> G is
the maximum genus among the genera of all surfaces in
which G has a 2-cell imbedding.

Duke [19] has shown the following:

Thm. 6-21. If a graph G has 2-cell imbeddings in S_m and S_n,
then G has a 2-cell imbedding in S_k,   for each
k,m ≤ k ≤ n.

Cor. 6-22. A connected graph G has a 2-cell imbedding in S_k if
and only if $\gamma(G) \le k \le \gamma_M(G)$.

An upper bound for $\gamma_M(G)$ is not difficult to determine.

Def. 6-23. The <u>Betti number</u>, $\beta(G)$, of a graph G having p
vertices, q edges, and k components, is given by:
$\beta(G) = q - p + k$.

$\beta(G)$ is sometimes called the <u>cycle rank</u> of G; it gives the
number of independent cycles in a cycle basis for G; see Harary
[28, pp. 37-40].
    Recall that [x] denotes the greatest integer less than or
equal to x; {x} gives the least integer greater than or equal to
x. Both symbols will be used frequently in the remainder of this
chapter.

Thm. 6-24.   Let   G   be connected; then $\gamma_M(G) \le \left\lceil \frac{\beta(G)}{2} \right\rceil$.   Moreover,

equality holds if and only if $r = 1$ or   2,   according
as $\beta(G)$ is even or odd, respectively.

Proof: Let   G   be connected, with a 2-cell imbedding in
S_k;   then $r \ge 1$,   and $\beta(G) = q - p + 1$;   also $p - q + r$
$= 2 - 2k$;   thus

$$k = 1 + \frac{q - p - r}{2} \le \frac{q - p + 1}{2} = \frac{\beta(G)}{2},$$

and the result follows.

Nordhaus, Stewart, and White [46] have shown that equality
holds in Theorem 6-24 for the complete graph K_n;   Ringeisen [50]
has shown that equality holds for the complete bipartite graph
$K_{m,n}$;   and Zaks [80] has shown that equality holds for the n-cube
Q_n:

Thm. 6-25. $\gamma_M(K_n) = \left\lceil \frac{(n-1)(n-2)}{4} \right\rceil$.

Thm 6-26. $\gamma_M(K_{m,n}) = \left\lceil \frac{(m-1)(n-1)}{2} \right\rceil$.

Thm. 6-27. $\gamma_M(Q_n) = (n-2)2^{n-2}$,   for $n \ge 2$.

Also, Ringeisen [51] has found $\gamma_M(G)$ for several classes of
planar graphs   G,   including the wheel graphs and the regular polyhedral graphs.

Nordhaus, Ringeisen, Stewart, and White have combined [45] to
establish the following analog to Kuratowski's Theorem (Theorem 6-6):
(The graphs   H   and   Q   are given in Figure 6-3.)

Thm. 6-28.   The connected graph   G   has maximum genus zero if and
only if it has no subgraph homeomorphic to either   H   or
Q.   (Furthermore, $\gamma(G) = \gamma_M(G)$ if and only if
$\gamma_M(G) = 0$ if and only if   G   is a cactus with vertex-
disjoint cycles.)

Def. 6-29.   A cactus is a connected (planar) graph in which every
block is a cycle or an edge.

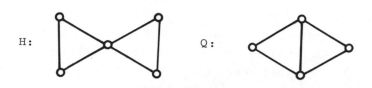

Figure 6-3

6-5.   Genus Formulae for Graphs

Theorems 6-25, 6-26, and 6-27, and the work of Ringeisen [51]
referred to above give the only known, non-trivial formulas for
maximum genus. Not very many more formulas are known for the genus
parameter; we list some of these below.

Thm. 6-30.   (Ringel [54]; Beineke and Harary [7])

$$\gamma(Q_n) = 1 + 2^{n-3}(n-4), \quad n \geq 2.$$

Thm. 6-31.   (Ringel [53])

$$\gamma(K_{m,n}) = \left\{\frac{(m-2)(n-2)}{4}\right\}; \quad m,n \geq 2.$$

Thm. 6-32.   (Ringel and Youngs [56])

$$\gamma(K_n) = \left\{\frac{(n-3)(n-4)}{12}\right\}, \quad n \geq 3.$$

Thm. 6-33.   (White [71]; see also [55], for the case   m = 1)

$$\gamma(K_{mn,n,n}) = \frac{(mn-2)(n-1)}{2}$$

Thm. 6-34.   (White [72]) Let G have p vertices of positive de-
             gree, q edges, k non-trivial components, and no 3-
             cycles. Let H have 2n (n ≥ 1) vertices and maximum
             degree less than two. Then $\gamma(G[H]) = k + n(nq-p)$.

Cor. 6-35.   Let G have no 3-cycles. Then $\gamma(G[K_2]) =$
             $\gamma(G[\overline{K}_2]) = \beta(G)$.

For a very readable proof of Theorem 6-30, see Behzad and Chartrand [4]. Theorems 6-31 and 6-33 are proved using schemes identical or similar to that discussed in the next section. Theorem 6-32 will be discussed at length in Chapter 9.

For an outline of the proof of Theorem 6-34, we offer the following: first we note that we may assume G to be connected (the general result will then follow from the Battle, Harary, Kodama, and Youngs Theorem). We then show that $G[H]$ can have at most $2p_Gq_H$ triangular regions in any imbedding (where G has order p_G and H has q_H edges), and proceed to construct an imbedding of $G[H]$ with $r_3 = 2p_Gq_H$ and $r_4 = r - r_3$. The euler formula shows that this imbedding will be minimal, and gives the desired formula for the genus. To construct the imbedding, we begin with q_G copies of $K_{2n,2n}$, each quadrilaterally imbedded in $S_{(n-1)^2}$ as described by Ringel (see Theorem 6-31). By making suitable vertex identifications (and this is the heart of the proof; see [72]), we obtain the graph $G[\overline{K_{2n}}]$, quadrilaterally imbedded in the desired surface. The construction thus far allows the q_H edges of each copy of H to be added; each such edge converts one quadrilateral region of the imbedding of $G[\overline{K_{2n}}]$ into two triangular regions. This gives the result.

For the above construction, one can compute the genus of the resulting surface directly, without recourse to any euler-type formula. The contributions to the genus are of three types:

(i) $q_G(n-1)^2$, representing the collective genera of the q_G 2-manifolds with which we began our construction;

(ii)   corresponding to every vertex v of G, we make $(\deg_G v-1)$ sets of $2n$ vertex identifications each, each set requiring a "bundle" of n tubes joining two 2-manifolds; this contributes $(2q_G-p_G)(n-1)$ to the genus;

(iii) $\beta(G) = q_G - p_G + 1$, representing the contribution of the bundles of tubes taken collectively.

Adding, we find:

$$\gamma(G[H]) = q_G(n-1)^2 + (2q_G-p_G)(n-1) + (q_G-p_G+1)$$

$$= 1 + n(nq_G-p_G).$$

6-6. Edmonds' Permutation Technique

Before leaving the theory of graph imbeddings and considering
specific imbedding problems, we present a powerful tool for solving
such problems: the Edmonds' permutation technique ([21]; see also
Youngs [77].) This amounts to an algebraic description, for any 2-
cell imbedding of a graph G. It is used, in one form or another,
in the proofs of the last four theorems listed above.

Denote the vertex set of a connected graph G by $V(G) =
\{1,2,\ldots,n\}$. For each $i \in V(G)$, let $V(i) = \{k \in V(G) \mid [i,k] \in E(G)\}$.
Let $p_i: V(i) \to V(i)$ be a cyclic permutation on $V(i)$, of length
$n_i = |V(i)|$. Then there is a one-to-one correspondence between 2-
cell imbeddings of G and choices of the p_i , given by:

Thm. 6-36. Each choice $(p_1,\ldots,p_n)$ determines a 2-cell imbedding
 $G(M)$ of G in a surface M, such that there is an
 orientation on M which induces a cyclic ordering of
 the edges [i,k] at i in which the immediate suc-
 cessor to [i,k] is $[i,p_i(k)]$, $i = 1,\ldots,n$. In fact,
 given $(p_1,\ldots,p_n)$, there is an algorithm which produces the determined imbedding. Conversely, given a
 2-cell imbedding $G(M)$ in a surface M with a given
 orientation, there is a corresponding $(p_1,\ldots,p_n)$ determining that imbedding.

 Proof: Let $D^* = \{(a,b) \mid [a,b] \in E(G)\}$, and define
 $P^*: D^* \to D^*$ by: $P^*(a,b) = (b,p_b(a))$. Then P^* is a
 permutation on the set D^* of directed edges of G
 (where each edge of G is associated with two oppo-
 sitely-directed directed edges), and the orbits under P^*
 determine the (2-cell) regions of the corresponding imbedding. These regions may then be "pasted" together --
 with (a,b) matched with (b,a) as in Figure 6-4 -- to
 form a surface M in which G is 2-cell imbedded.
 (Since every edge (a,b) in the boundary of a given region is matched with an edge -- (b,a) -- in the
 boundary of another (or possibly the same) region, M is
 closed. Since (a,b) is matched with (b,a) -- and
 not with (a,b) -- M is orientable. Since each p_i

is a <u>cyclic</u> permutation, M is a 2-manifold.) The
genus of M may now be determined by the euler formula,
with r given by the number of orbits under P. The
converse follows from similar considerations.

As an example, consider the imbedding of $K_{3,3}$ in S_1 depicted in Figure 6-5. Let $V(K_{3,3}) = \{1,2,3,4,5,6\}$, with
$V(1) = V(2) = V(3) = \{4,5,6\}$; $V(4) = V(5) = V(6) = \{1,2,3\}$. Then

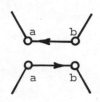
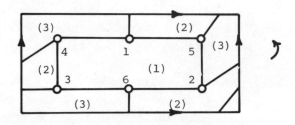

Figure 6-4   Figure 6-5

$$p_1: (4,5,6) \qquad p_4: (1,2,3)$$
$$p_2: (4,5,6) \qquad p_5: (1,2,3)$$
$$p_3: (4,5,6) \qquad p_6: (1,2,3)$$

describe this imbedding. The orbits under P^* are:

(1)   (1,5) (5,2) (2,6) (6,3) (3,4) (4,1)

(2)   (5,1) (1,6) (6,2) (2,4) (4,3) (3,5)

(3)   (2,5) (5,3) (3,6) (6,1) (1,4) (4,2)

(Note that $P^*(4,1) = (1,5)$; $P^*(3,5) = (5,1)$; $P^*(4,2) = (2,5)$.)
    As a matter of notation, from this point on, we will abbreviate
an orbit such as (1) above by: 1-5-2-6-3-4; it is implicit that
$p_4(3) = 1$, and $p_1(4) = 5$.
    It now follows that the genus of any connected graph (and
hence, by Corollary 6-19, of any graph) can be computed, by selecting, from among the $\prod_{i=1}^{n} (n_i-1)!$ possible permutations P^*, one

which gives the maximum number of orbits, and hence determines the
genus of the graph (component).   (Since, by Theorem 6-11, a minimal
imbedding must be a 2-cell imbedding, it corresponds to some P^*;
by Corollary 5-15, r will be minimal for this imbedding.)   The
obvious difficulty in applying this procedure is that of selecting
a suitable P^* from the (usually) vast number of possible ordered
n-tuples of local vertex permutations P_i.

6-7.   Imbedding Graphs on Pseudosurfaces

In Section 5-5 we introduced the pseudosurfaces $S(k;n_1(m_1)$,
$...,n_t(m_t))$. Recall that for any imbedding of a graph G in a
pseudosurface S', we assume that each singular point of S' is
occupied by a (singular) vertex of G.   The number $2 - 2k - \sum_{i=1}^{t} n_i(m_i-1)$ gives the characteristic of S', denoted by $\chi'(S')$.

Def. 6-37.   The <u>pseudocharacteristic</u>, $\chi'(G)$, <u>of a graph</u>
            G is the largest integer $\chi'(S')$ for all pseudo-
            surfaces S' in which G can be imbedded.

A surface can be considered as a (degenerate) pseudosurface.
Hence we have

$$\chi'(G) \geq 2 - 2k\gamma(G).$$

Equality may be strict (i.e. pseudosurfaces may be more efficient,
from the point of view of maximizing characteristic, for imbedding
graphs into); for example $\chi'(K_5) = 1$, as Figure 6-6 shows.

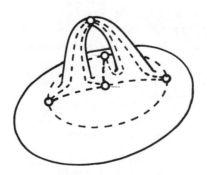

Figure 6-6

Petroelje [49] has found that many of the basic theorems for imbedding graphs in surfaces carry over for pseudosurfaces. For example:

Thm. 6-38. Let G be a connected graph minimally imbedded in the pseudosurface S'; then the imbedding must be 2-cell.

Thm. 6-39. If G(p,q) is connected, then $\chi'(G) \leq p-q/3$; equality holds if and only if G has a triangular imbedding in some pseudosurface.

Petroelje also developed an analogue of Edmonds' permutation technique for pseudosurfaces, and found the following formulae (among others):

Thm. 6-40. $\chi'(K_{n,n,n,n}) = 2n(2-n)$.

(This is consistent with the conjecture of Ringel that $\gamma(K_{n,n,n,n}) = (n-1)^2$, with $r = r_3$; see Problem 9-10.)

Thm. 6-41. $\chi'(K_{2m,2n,r}) = 2(m+n-mn) - r(m-1)$, where $2m \geq 2n \geq r \geq 1$.

6-8. Other Topological Parameters for Graphs

We have seen that, if a graph is not planar, we can still make a "proper" drawing of the graph in some surface and/or pseudo-surface. Two common topological parameters (other than genus) which arise if modified drawings are allowed are the thickness and crossing number.

Def. 6-42. The thickness $\theta(G)$ of a graph G is the minimum number of planar subgraphs whose union is G.

For sample formulae (see [6], [5], and [33], respectively) we have:

Thm. 6-43. For $n \not\equiv 4 \pmod 6$, $n \neq 9$,

$$\theta(K_n) = \left\lceil \frac{n+7}{6} \right\rceil .$$

<u>Thm. 6-44</u>. $\theta(K_{n,n}) = \left\lceil \dfrac{n+5}{4} \right\rceil$.

<u>Thm. 6-45</u>. $\theta(Q_n) = \left\{ \dfrac{n+1}{4} \right\}$.

<u>Def. 6-46</u>.   The <u>crossing number</u> $v(G)$ of a graph G is the minimum
number of pairwise intersections of its (open) edges,
among all drawings of G in the plane.

One might say that the crossing number tells us, if we insist
upon drawing G on S_0,   just how bad this drawing must be.   For
this parameter, exact values are scarce; we mention the following
(see [27]):

<u>Thm. 6-47</u>. $v(K_n) \le \dfrac{1}{4}\left\lfloor\dfrac{n}{2}\right\rfloor\left\lfloor\dfrac{n-1}{2}\right\rfloor\left\lfloor\dfrac{n-2}{2}\right\rfloor\left\lfloor\dfrac{n-3}{2}\right\rfloor$;   equality holds for
$n \le 10$.

<u>Thm. 6-48</u>. $v(K_{m,n}) \le \left\lfloor\dfrac{m}{2}\right\rfloor\left\lfloor\dfrac{m-1}{2}\right\rfloor\left\lfloor\dfrac{n}{2}\right\rfloor\left\lfloor\dfrac{n-1}{2}\right\rfloor$;   equality holds for $m \le 6$.

As an indication of the kind of techniques that might be employed, we prove the following:

<u>Thm. 6-49</u>. $v(K_{3,2,2}) = 2$.

> <u>Proof</u>: Suppose $v(K_{3,2,2}) = x$.   Since $K_{3,3}$ is a sub-
> graph of $K_{3,2,2}$, $\gamma(K_{3,2,2}) \ge 1$ (by the hint for
> Problem 6-3).   Thus $x \ge 1$.   Consider such an optimal
> drawing of $K_{3,2,2}$ in S_0; this gives rise to a plane
> graph G, with $p = 7 + x$, $q = 16 + 2x$,   and $r \ne r_3$.
> (If $r = r_3$ the configuration in Figure 6-7a, which
> must exist since $x \ge 1$,   must correspond to the con-
> figuration in Figure 6-7b, which cannot occur in any
> complete tripartite graph.)   Hence $2q = 32 + 4x \ge$

                (a)   (b)

Figure 6-7

$4 + 3(r-1) = 3r + 1$, and $9 + x = q - p = r - 2 \leq$
$\frac{31}{3} + \frac{4}{3}x - 2 = \frac{25}{3} + \frac{4}{3}x$. Thus $\frac{2}{3} \leq \frac{1}{3}x$, so that $x \geq 2$.
Figure 6-8 shows that $x \leq 2$, to complete the proof.

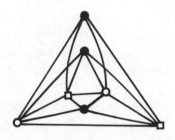

Figure 6-8

A natural extension of the construction in Figure 6-8 gives the following:

Thm. 6-50. $v(K_{m,n,r}) \leq f(m,n) + f(m,r) + f(n,r)$, where $f(x,y) =$
$(x-1)\binom{y-1}{2} + (y-1)\binom{x-1}{2} + \binom{x-1}{2}\binom{y-1}{2} = \frac{(x-1)(y-1)(xy-4)}{4}$.

Cor. 6-51. $v(K_{n,n,n}) \leq \frac{3}{4}(n-2)(n-1)^2(n+2)$; equality holds for $n = 1,2$.

6-9. Applications

For applications of the four topological parameters discussed in this chapter, consider the problem of printing an electronic circuit on a circuit board. If the associated graph G is planar, one board will suffice, without modification. If G is not planar, at least four alternatives are available to avoid short circuits (the choice depending upon relevant considerations of an engineering and/or economic nature): 1.) the circuit can be accommodated by drilling holes through the board; $\gamma(G)$ gives the minimum number of holes; 2.) some of the vertices can be printed on both sides of the board, with connections made through the board between corresponding images of the same vertex; here we seek the minimum n such that G can be imbedded in the pseudosurface $S(0; n(2))$; however, it may occur that no such n exists (see Problem 6-7.) If a given vertex can appear arbitrarily often, with connections made through the

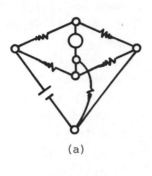

(a)

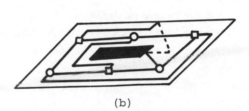

(b)

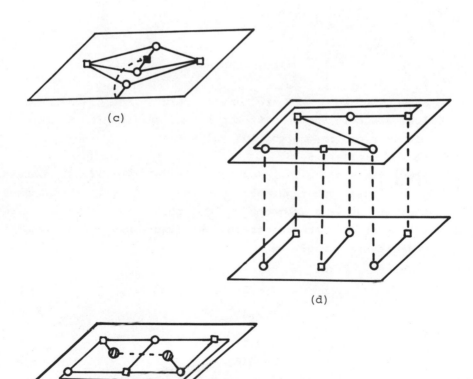

(c)

(d)

(e)

Figure 6-9

board among corresponding images of the same vertex, and if in additional holes can be drilled as in 1.), then we seek the value of the
parameter $\chi'(G)$, for maximum efficiency; 3.) If several circuit
boards are used, each containing a planar portion of the circuit,
and jumpers are run between successive boards to connect corresponding images of the same junction, then we are studying the parameter $\theta(G)$; 4.) If the circuit is stamped on one side of one circuit board (with no holes yet drilled) and if wherever two connections cross extraneously two holes are now drilled to allow one
connection to temporarily pass to the other side of the board, en-
abling it to "cross" the second connection while avoiding a short
circuit, it is the parameter $v(G)$ that dictates economy of effort
here.

As an example, consider the modified wheatstone bridge circuit
of Figure 6-9(a); the associated graph is $G = K_{3,3}$. Figures 6-9(b)-
(e) correspond respectively to: $\gamma(K_{3,3}) = 1$, $\chi'(K_{3,3}) = 1$,
$\theta(K_{3,3}) = 2$, and $v(K_{3,3}) = 1$.

6-10. Problems

6-1.)   Show that two graphs are homeomorphic in the graph-theoretical sense if and only if their realizations in E^3 are
homeomorphic in the topological sense.

6-2.)   Prove Corollary 6-15.

6-3.)   Prove the easy half of Kuratowski's Theorem: if G contains
a Kuratowski subgraph, then G is non-planar. (Hint: show
that if H is a subgraph of G, then $\gamma(H) \le \gamma(G)$.)

6-4.)   Show that the Petersen graph (see Figure 8-9) is non-planar.
What is its genus?

6-5.)   Show that p_1, p_3: (5,6,7,8)

$\qquad p_2, p_4$: (8,7,6,5)

$\qquad p_5, p_7$: (1,2,3,4)

$\qquad p_6, p_8$: (4,3,2,1)

describe a 2-cell imbedding of $K_{4,4}$ in S_1, with $r = r_4$.

6-6.)   Why is $\chi'(G)$ defined as a maximum characteristic, instead
of a minimum genus?

6-7.)   Show that $G = K_n$, $n \ge 13$, imbeds on no pseudosurface
$S(0; k(2))$.

Chapter 7.   The Genus of a Group

To get an accurate and efficient "picture" of a group, we seek
a surface of minimum genus on which we can imbed a Cayley color
graph of some presentation of the group. This suggests the following definition; let $\gamma(D_p(\Gamma))$ denote the genus of the underlying
graph $G_\Delta(\Gamma)$ determined from $D_p(\Gamma)$ by removing all arrows and
colors from the edges (recall that, by convention, $D_p(\Gamma)$ has no
multiple edges). Then:

Def. 7-1. The genus of a group Γ is given by:

$$\gamma(\Gamma) = \min\{\gamma(D_p(\Gamma))\},$$

where the minimum is taken over all presentations   P
for Γ.

Def. 7-2. A group Γ is said to be planar if $\gamma(\Gamma) = 0$.

7-1.   Imbeddings of Cayley Color Graphs

Finite planar groups have been cataloged by Maschke [41] (see
also Anderson [1]). The finite planar groups on one generator are
exactly the cyclic groups Z_n; on two generators, they include the
dihedral groups D_n, groups of the form $Z_2 \times Z_n$, S_4, A_4, and
A_5 (the last three groups are the symmetry groups of the regular
polyhedra), and $Z_2 \times A_4$; on three generators (each must be of
order 2) finite planar groups include $Z_2 \times D_n$, $Z_2 \times S_4$,   and

$Z_2 \times A_5$. In summary:

Thm. 7-3. The finite group G is planar if and only if $G = G_1 \times G_2$,
          where $G_1 = Z_1$ or Z_2 and $G_2 = Z_n$, D_n, S_4, A_4, or A_5.

We consider infinite groups temporarily, preparatory to establishing a startling result, due to Levinson [37]. In this chapter,
an infinite graph is given by:

Def. 7-4. An _infinite graph_ is a graph with denumerable vertex set.

There are two natural (but non-equivalent!) definitions of
planarity, for infinite graphs.

Def. 7-5. An infinite graph is said to be _planar_ if it can be im-
          bedded in the plane.

)ef. 7-5'. An infinite graph is said to be _planar_ if it can be
           imbedded in the plane so that the vertex set has no
           limit points.

We adopt Definition 7-5, for reasons soon to be obvious.

Thm. 7-6. An infinite graph is planar if and only if it contains
          no subgraph homeomorphic with K_5 or $K_{3,3}$.

For a proof of this extension of Kuratowski's theorem, see
Dirac and Schuster [18]. To see that this extension does _not_ hold
for Definition 7.5', consider the graph of Figure 7-1, where an
infinite path is attached at each vertex of K_4.

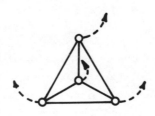

Figure 7-1

<u>Def. 7-7</u>.   An infinite graph G has <u>infinite genus</u> ($\gamma(G) = \infty$), if,
            for every natural number n, there exists a finite sub-
            graph G_n of G such that $\gamma(G_n) \geq n$.

<u>Lemma 7-8</u>.   Let G be the graph of a presentation of an infinite
            group Γ. Let H be an induced finite subgraph of G.
            Then there exist two disjoint, isomorphic copies of H
            in G.

            <u>Proof</u>: The vertex set H corresponds to a finite set
            $\{g_1, \ldots, g_n\}$ of elements of Γ. Form the (finite) set:

            $$S = \{g_i g_j^{-1} \mid 1 \leq i, \ j \leq n\}.$$

            Pick $x \in \Gamma - S$. Form H^*, the subgraph of G induced
            by $\{xg_i \mid i = 1, \ldots, n\}$; then H^* is isomorphic to H,
            since $g_i h = g_j$ if and only if $xg_i h = xg_j$. Now, sup-
            pose that $v \in V(H) \cap V(H^*)$; then there exist i and
            j such that $xg_j = g_i$,   so that $x = g_i g_j^{-1} \in S$, a
            contradiction.

    In Figure 7-2, portions of Cayley color graphs for presentations
of three infinite groups are given. The second group (b) is
called the <u>infinite dihedral</u> group; the third group (c) is the
<u>free group</u> on two generators.
    We now present Levinson's result:

<u>Thm. 7-9</u>.   Let Γ be a countably infinite group, with G the graph
            of a presentation for Γ. Then either $\gamma(G) = 0$, or
            $\gamma(G) = \infty$.

            <u>Proof</u>: Suppose G is not planar; then, by Theorem 7-6,
            G contains K, a Kuratowski (and hence finite) sub-
            graph. Thus $\gamma(G) \geq 1$. Let n be an arbitrary natural
            number. But by Lemma 7-8, we can find a second, disjoint
            copy of K in G, so that $\gamma(G) \geq \gamma(2K) = 2$ by
            Corollary 6-19. Now apply the lemma again, with H = 2K,
            to obtain two disjoint copies of 2K in G, so that
            $\gamma(G) \geq 4$. Continuing in this fashion, we eventually find
            two disjoint copies of $2^{n-1}K$ in G, so that $\gamma(G) \geq$
            $2^n > n$; then $\gamma(G) = \infty$, since n was arbitrary.

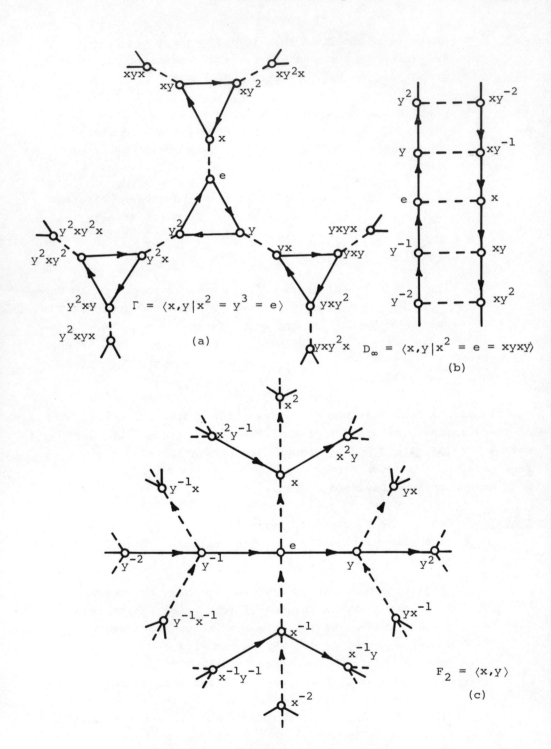

Figure 7-2

Returning our attention to finite groups, we produce examples of groups of positive genus. The following lemma will be useful. Note that if P gives $\gamma(\Gamma)$, then P may be assumed to have no redundant generators; i.e. P is minimal. We also note that, in any imbedding of a Cayley Color graph, every region boundary corresponds to an identity word.

<u>Lemma 7-10</u>. Let Γ be a finite group, with $3 \nmid |\Gamma|$; let P be a minimal presentation for Γ. Then $D_{P.}(\Gamma)$ contains no triangles.

   <u>Proof</u>: Suppose $D_P(\Gamma)$ contains a triangle; then we find a closed walk $h_1^{a_1} h_2^{a_2} h_3^{a_3} = e$ in $D_P(\Gamma)$, where h_i is a generator in P, and $a_i = \pm 1$. If any two of the h_i are distinct, then one of these two is redundant. If, on the other hand, $h_1 = h_2 = h_3$, then the a_i all have the same sign (or else all three $= e$). But then $h_1^3 = e$, and $3 \mid |\Gamma|$, a contradiction.

Now consider Q, the group of the quaternions. Let P be a presentation for Q, such that $\gamma(Q) = \gamma(D_P(Q))$; then P is minimal. By Lemma 7-10, $D_P(Q)$ has no triangles, since $|Q| = 8$. It is not difficult to see that P has at least two generators and that if P has exactly two generators, neither can be of order 2; furthermore, P cannot have three generators of order 2 (see Problem 7-1). Thus $D_P(Q)$ is regular of degree at least four. Then $2q \geq 4p = 32$; i.e. $q \geq 16$. Now, by Corollary 6-15,

$$\gamma(Q) \geq \frac{q}{4} - \frac{p}{2} + 1 \geq 1.$$

That $\gamma(Q) = 1$ is shown by Figure 7-3, where $Q = \langle x,y \mid x^2 = y^2 = (xy)^2 \rangle$.

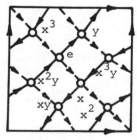

Figure 7-3

7-2.   Genus Formulae for Groups

We now find a non-trivial genus formula for an infinite class of groups: those groups (necessarily abelian) in which every element is of order 2. Let Γ_n denote this group; then $\Gamma_n = (Z_2)^n$, and $|\Gamma_n| = 2^n$.

__Thm. 7-11.__ $\gamma(\Gamma_n) = 1 + 2^{n-3}(n-4)$, $n \geq 2$.

> __Proof:__ Γ_n may be expressed as follows: $\Gamma_1 = Z_2$; $\Gamma_n = Z_2 \times \Gamma_{n-1}$, for $n \geq 2$. Writing Γ_n as an iterated direct product in this way, we see that any P for Γ_n must have at least n generators; hence $2q \geq np = n2^n$; thus by Lemma 7-10 and Corollary 6-15,
>
> $$\gamma(\Gamma_n) \geq \frac{q}{4} - \frac{p}{2} + 1$$
>
> $$\geq n2^{n-3} - 2^{n-1} + 1$$
>
> $$= 1 + 2^{n-3}(n-4).$$
>
> But now let P be determined by repeated application of Theorem 4-16; then $G_P(\Gamma_n) = Q_n$, the n-cube, and
>
> $$\gamma(\Gamma_n) \leq \gamma(D_P(\Gamma_n))$$
>
> $$= \gamma(Q_n)$$
>
> $$= 1 + 2^{n-3}(n-4),$$
>
> by Theorem 6-30.   This completes the proof.

Let us extend this result somewhat. We will need the following genus formula, involving $(n+1)$ parameters. Define the graph H_n as follows: let $H_1 = C_{2m_1}$, the cycle on $2m_1$ vertices, and recursively define $H_n = H_{n-1} \times C_{2m_n}$, for $n \geq 2$, where each $m_i \geq 2$. Let $M^{(n)} = \prod_{i=1}^{n} m_i$.

__Thm. 7-12.__ $\gamma(H_n) = 1 + 2^{n-2}(n-2)M^{(n)}$, $n \geq 2$.

> __Proof:__ By Theorem 2-19, Problem 1-4, and a trivial induction argument, H_n is a bipartite graph. We produce a quadrilateral imbedding for H_n, and compute $\gamma(G_n)$

using Corollary 6-15. For H_n, let $p^{(n)}$ and $q^{(n)}$ denote the number of vertices and edges respectively. Then $p^{(n)} = 2^n M^{(n)}$; and since H_n is regular of degree $2n$, it is a simple matter to compute $q^{(n)} = 2^n n M^{(n)}$.

Let the statement $S(n)$ be: there is an imbedding of H_n for which $r = r_4 = n2^{n-1} M^{(n)}$, including two disjoint sets of $2^{n-2} M^{(n)}$ mutually vertex-disjoint quadrilateral regions each, both sets containing all $2^n M^{(n)}$ vertices of H_n. We claim that $S(n)$ is true for all $n \geq 2$, and we verify this claim by mathematical induction.

That $S(2)$ is true is apparent from Figure 7-4 (which shows an imbedding of $C_4 \times C_6$ in S_1), with the regions designated by (1) making up one set, and those designated by (2) making up the other. We now assume $S(n)$ to be true and establish $S(n+1)$, for $n \geq 2$.

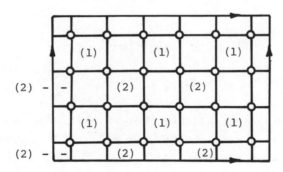

Figure 7-4

For the graph H_{n+1}, we start with $2m_{n+1}$ copies of H_n, minimally imbedded as described by $S(n)$. We partition the corresponding surfaces into m_{n+1} copies of one orientation, and m_{n+1} copies of the reverse orientation, corresponding to the vertex set partition of the bipartite graph $C_{2m_{n+1}}$. From each copy, two joins of $p^{(n)}$ edges each must be made, both to copies of opposite orientation, in order to construct H_{n+1}. From the statement $S(n)$, it is clear that these two joins can be made, each one over $2^{n-2} M^{(n)}$ tubes

carrying four edges each. (Attach one end of a tube in
the interior of each region designated by (1) for one
join; use the regions designated by (2) for the second
join.) Each new region formed by this process is a
quadrilateral. In this fashion the required $2m_{n+1}$
joins can be made to imbed H_{n+1}, with $r = r_4$. Now
form one set of regions by selecting opposite quadrilaterals from each tube added in alternate joins in this
construction. Form the second set by selecting the remaining quadrilaterals on the same tubes. It is clear
that the two sets of regions thus selected are disjoint,
and that each contains $(2)(m_{n+1})(2^{n-2}M^{(n)}) = 2^{n-1}M^{(n+1)}$
mutually vertex-disjoint quadrilaterals; both sets contain all $2^{n+1}M^{(n+1)}$ vertices of H_{n+1}. Furthermore,
$r^{(n+1)} = 2m_{n+1}r^{(n)} + \Delta r$, where $\Delta r = (2m_{n+1})(2^{n-2}M^{(n)})(2)$,
where $2m_{n+1}$ joins have been made, with $2^{n-2}M^{(n)}$ tubes
per join, and a net increase in r of 2 per tube.
Hence,

$$r^{(n+1)} = 2m_{n+1}(n2^{n-1}M^{(n)}) + 2^nM^{(n+1)}$$

$$= (n+1)2^nM^{(n+1)},$$

and we have established that $S(n+1)$ follows from $S(n)$.
Therefore, $S(n)$ holds, for all $n \geq 2$.

We can now compute:

$$\gamma(H_n) = 1 + \frac{2^n nM^{(n)}}{4} - \frac{2^nM^{(n)}}{2}$$

$$= 1 + 2^{n-2}(n-2)M^{(n)}.$$

For the special case where $m_i = m$, $i = 1, \ldots, n$, we have
$M^{(n)} = m^n$, and:

<u>Cor. 7-13</u>. The genus of $H_n^{(m)}$ is given by:

$$\gamma(H_n^{(m)}) = 1 + 2^{n-2}(n-2)m^n.$$

Furthermore, if $m = 2$ in the above formula, since $C_4 = K_2 \times K_2$,

$H_n^{(2)}$ is the 2n-cube, and we obtain the result (compare with Theorem 6-30):

Cor. 7-14. $\gamma(Q_{2n}) = 1 + 2^{2n-2}(n-2)$.

For further results concerning the genus of repeated cartesian products of bipartite graphs, see [73].

Now, let $\Gamma_n^{(m)}$ be the abelian group with minimal Cayley color graph $H_n^{(m)}$, $m \geq 2$; i.e. $\Gamma_1^{(m)} = Z_{2m}$, and $\Gamma_n^{(m)} = Z_{2m} \times \Gamma_{n-1}^{(m)}$, for $n \geq 2$. Then we have:

Cor. 7-15. $\gamma(\Gamma_n^{(m)}) = 1 + 2^{n-2}(n-2)m^n$.

The reader may wish to combine Theorems 4-15, 4-18, and 7-12 to obtain genus formulae for additional abelian groups. For example, using the notation of Theorem 4-15, consider Γ where m_r is even. The argument of Theorem 7-11 can be modified to assist in the computation of the genus for certain hamiltonian groups; the following results are due to Himelwright [31]:

Thm. 7-16. $\gamma(Q \times (Z_2)^n) = n\,2^n + 1$.

Thm. 7-17. $\gamma(Q \times Z_n \times (Z_2)^n) = mn2^n + 1$,   for   m   odd.

Cor. 7-18.   The groups $Q \times Z_m \times (Z_2)^8$,   for   m   odd,   have genus asymptotic to the order.

By Theorems 4-15 and 4-21, if G is a hamiltonian group, then $G = Q \times Z_{m_1} \times \cdots \times Z_{m_r} \times (Z_2)^n$, where the m_i are odd (i = 1, ..., r) and $m_i | m_{i-1}$ (i = 2, ..., r). Himelwright has also shown:

Thm. 7-19.   The genus of the hamiltonian group $Q \times Z_{m_1} \times \cdots \times Z_{m_r} \times (Z_2)^n$ is asymptotic to $2^n(r+n-1)\prod_{i=1}^{r} \dfrac{r}{\pi}$,   if $1 \leq r \leq n-1$.

There are many open questions in this area. If a generalization of Theorem 7-12 for products of arbitrary (not necessarily even) cycles could be found, then the genus of any abelian (and also of any hamiltonian) group could be easily computed. What is $\gamma(S_n)$?

$\gamma(A_n)$? The following theorem produces upper bounds for these group genera.

<u>Thm. 7-20</u>. If Γ is minimally generated by $\{g_1,\ldots,g_n\}$ and

satisfies at least the relations $g^{m_i} = e = (\overset{n}{\underset{j=1}{\pi}} g_j)^k$,

then

$$\gamma(\Gamma) \leq 1 + \frac{|\Gamma|}{2}\left(n-1 - \frac{1}{k} - \sum_{j=1}^{n} \frac{1}{m_j}\right) .$$

<u>Proof</u>: Select $p_g = (gg_1, gg_1^{-1}, gg_2, gg_2^{-1}, \ldots, gg_n, gg_n^{-1})$, for all $g \in G$. Then, using Edmond's algorithm (see Theorem 6-33), we compute orbits as follows:

(i)   An orbit containing the directed edge (a, ag_i^{-1}) continues with $p_{ag_i^{-1}}(a) = ag_i^{-2}$; hence this orbit corresponds to the relation $g_i^{m_i} = e$ and has length m_i. (If $m_i = 2$, we draw edges for both $gg_i = g'$ and $g'g_i = g$, obtaining $\frac{|\Gamma|}{2}$ 2-sided regions; for each such region, the two sides may be identified and the arrows removed, so that the region is destroyed but the genus is unaffected.)

(ii) An orbit containg the directed edge (a, ag_i) continues with $p_{ag_i}(a) = ag_i g_{i+1}$; hence this orbit corresponds to the relation $(\overset{n}{\underset{j=1}{\pi}} g_j)^k = e$ and has length nk. As there are no other orbits, we find:

$$r = \sum_{i=1}^{n} r_{m_i} + r_{nk}$$

$$= \sum_{i=1}^{n} \frac{|\Gamma|}{m_i} + \frac{|\Gamma|}{k};$$

the euler formula now gives the genus γ of the theorem for this imbedding of $D_P(\Gamma)$, for this presentation P for Γ. Hence $\gamma(\Gamma) \leq \gamma(D_P(\Gamma)) \leq \gamma$.

We note that an equivalent formula was obtained by Burnside [10, p. 398] in a different context. Theorem 7-20 gives $\gamma(G)$ exactly, for $\Gamma = Z_m$, A_4, S_4, A_5, or $Z_3 \times Z_3$. We also obtain the following two corollaries:

<u>Cor. 7-21</u>. $\gamma(S_n) \leq 1 + \frac{(n-2)!}{3}(n^2-5n+2)$, $n \geq 2$.

       <u>Proof</u>: Take $s = (1\,2\,3\,\ldots\,n)$ and $t = (1\,2)$ as generators for S_n; then $s^n = t^2 = (st)^{n-1} = e$.

<u>Cor. 7-22</u>.

$$\gamma(A_n) \leq \begin{cases} 1 + \dfrac{(n-1)(n-3)!}{8}(n^2-6n+4)\,, & n \quad \text{odd} \\[3ex] 1 + \dfrac{n(n-2)!(n-5)}{8}\,, & n \quad \text{even.} \end{cases}$$

       <u>Proof</u>: For n odd, $s = (1\,2\,\ldots\,n-2)$ and $t = (1\,n-1)(2\,n)$ generate A_n, and $s^{n-2} = t^2 = (st)^n = e$. For n even, $s = (1\,2\,\ldots\,n-1)(2\,n)$ and $t = (1\,2)(3\,n)$ generate A_n, with $s^{n-1} = t^2 = (st)^{n-1} = e$ (see [9]).

The two formulas given above for S_n and A_n respectively were also found by Brahana [9], using a different method and in a slightly different context. For related results, see [70].

7-3. Problems

7-1.) Show: that any presentation P for Q, the quaternions, has at least two generators; that if P has exactly two generators, neither can be of order 2; and that P can not have exactly three generators, each of order 2. (Hence $\delta(D_P(Q)) \geq 4$).

7-2.) Find an example of a group Γ and a presentation P for Γ such that $\gamma(D_P(\Gamma)) = \infty$.

7-3.) Find $\gamma(Z_m \times Z_n)$, for all m and n.

7-4.) Show that the only finite planar abelian groups are Z_n, $Z_2 \times Z_{2n}$, and $Z_2 \times Z_2 \times Z_2$, where $n \geq 1$.

*7-5.)   Use the imbedding of Problem 6-5 to find $\gamma(Q \times Q)$.

*7-6.)   Show that the <u>dicyclic group</u> $G_n = \langle x,y \mid x^{2n} = x^n y^{-2} = y^{-1}xyx = e \rangle$ has genus 1, for all $n > 1$. $(G_2 = Q;$ G_3
         is the "least familiar group of order 12.")

**7-7.) If Γ_1 is a subgroup of Γ_2,   is $\gamma(\Gamma_1) \leq \gamma(\Gamma_2)$? Prove or
         disprove!

Chapter 8: Map-coloring Problems

   In this chapter we will see that the famous four-color conjecture can be formulated -- and studied -- in graph-theoretical terms.
Graph theory will be used to establish the five-color theorem. The
Heawood Map-coloring theorem will be introduced; this powerful
theorem, whose proof was completed in 1968, answers the coloring
question -- which is still unanswered for the sphere -- for every
other closed 2-manifold. The easy half of the proof -- found by
Heawood in 1890 -- is presented in this section. The difficult half
of the proof -- developed primarily by Ringel and Youngs -- will be
discussed in Chapter 9.
   Consider any map of the world. Suppose we desire to color
the countries of the world (or the states of a particular country,
or the counties of a particular state, etc.) so that distinct
countries are distinguishable. This means that if two countries
share a border at other than isolated points, then they must be
colored differently. We make only one assumption as to the
countries themselves: each country must be connected (this rules out
Pakistan of the last decade, and the United States, for example.)
Note that a country need not be a 2-cell; that is, it may entirely surround some collection of other countries (such is the case
for a certain region in France; see Fréchet and Fan [23], p. 3).
   We mention in passing that several generalizations of this
map-coloring problem are possible. One of the most appealing is the
following: allow disconnected countries, with each country having at
most k components. (It is not hard to see that, without this restriction involving k, arbitrarily many colors may be needed.)
Then it can be shown (see Problem 8-6 for the case k = 2) that 6k
colors will always suffice. Ringel ([52]; p. 26) has found a map,
for the case k = 2, requiring 12 colors, so that this case is
completely solved.
   Returning now to the case of classical interest (k = 1), we
pose the question thusly: what is the smallest number of colors
needed to color any map on the sphere (or, equivalently, on the
plane)? That four colors may be needed is indicated by the map in-

duced by the tetrahedron. That five colors suffice for the sphere
will be demonstrated shortly. Whether or not five colors are ever
necessary has probably stimulated as much work in mathematics as any
otner single mathematical question; but the answer remains unknown.
The four-color conjecture says that five colors are never necessary;
four colors will suffice to color any map on the sphere. Many
"proofs" of the four color conjecture have been presented to the
mathematical community, but none has yet survived close scrutiny.
The interested reader might wish to read through one of these,
given by Kempe in 1879 (see [4], for example), and try to spot the
error in the "proof."

   Graph theory enters the picture in the following way. Form the
dual of the map in question. This produces a pseudograph. Attempt
to color tne vertices of the pseudograph so that no two adjacent
vertices have the same color. The pseudograph has no loops, as no
self-respecting country ever shares a border with itself. In fact,
we may as well drop any multiple edges, since they (the "extra"
edges) have no bearing on the coloring question. Then the coloring
numbers, or chromatic numbers, of the resulting graph and the map
will be identical. This leads to the following definitions.

8-1. Definitions

Def. 8-1.   The chromatic number, $\chi(G)$, of a graph G is the
            smallest number of colors for $V(G)$ so that adjacent
            vertices are colored differently.

Def. 8-2.   The chromatic number, $\chi(S_k)$, of a surface S_k is the
            largest $\chi(G)$ such that G can be imbedded in S_k.

8-2. The Four-color Conjecture

In this terminology, the four-color conjecture becomes:

Conjecture 8-3. $\chi(S_0) = 4$.

   There are many other equivalent formulations of the four-color
conjecture (see, for example, Ore's book: The Four Color Problem
[47], or [4]). We consider two of these.

Thm. 8-4.   The four-color conjecture is true if and only if every
cubic plane block is 4-region colorable.

Proof: Clearly if every plane graph is 4-region colorable, so is every cubic plane block. For the converse,
assume every cubic plane block is 4-region colorable, and
let G be a plane block. We obtain a cubic plane block
G' from G, by repeated operations of the form (a) --
for vertices of degree 2 -- and (b) -- for vertices of
degree 3 or more -- as depicted in Figure 8-1. Thus
G' is 4-region colorable (by hypothesis), and any

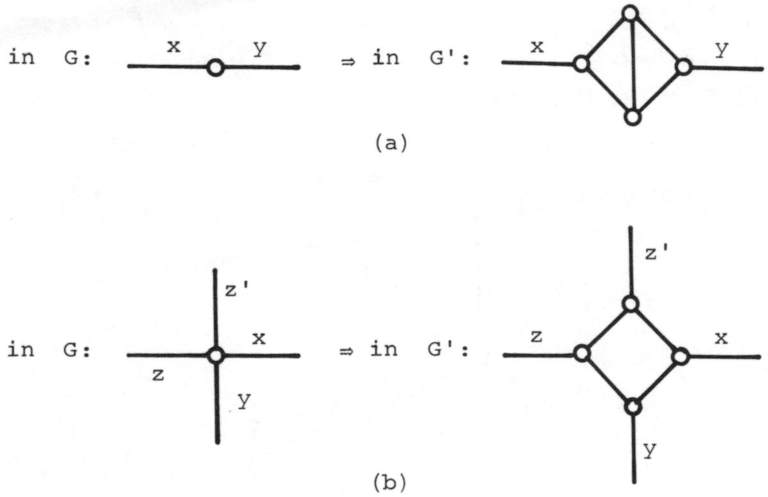

(a)

(b)

Figure 8-1

4-region coloring of G' induces a 4-region coloring for
G. Since it is apparent that the region coloring number
of an arbitrary plane graph is the maximum of the corresponding numbers for its blocks, the proof is complete.

Def. 8-5.   A graph G is said to be n-edge colorable if colors can
be assigned to E(G) so that adjacent edges are colored
differently.

<u>Thm. 8-6</u>.   The Four-Color Conjecture is true if and only if every
cubic plane block is 3-edge colorable.

<u>Proof</u>:   By Theorem 8-4, it suffices to show that a cubic
plane block   G   is 4-region colorable if and only if it
is 3-edge colorable.

  Suppose   G   is 4-region colorable, and let the
colors be taken from the group $\Gamma = Z_2 \times Z_2$.   Since   G
is a block, each edge   x   of   G   appears in the boundary
of two distinct (but adjacent) regions, R_x^1 and R_x^2.
Define the color of   x   by $c(x) = c(R_x^1) + c(R_x^2)$,   addition taking place in Γ.   Since $c(R_x^1) \neq c(R_x^2)$, $c(x) \neq e$,
the identity of Γ (every element is its own inverse, in
Γ.)   Let   x, y,   and   z   be adjacent edges in   G;   see
Figure 8-2.   We claim that   x, y,   and   z   are colored

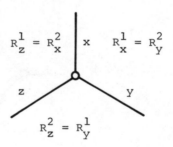

Figure 8-2

distinctly.   Suppose to the contrary that, say, $c(x) =$
$c(y)$;   that is $c(R_x^1) + c(R_x^2) = c(R_y^1) + c(R_y^2) = c(R_y^1) +$
$c(R_x^1)$.   But then $c(R_y^1) = c(R_x^2)$,   a contradiction.   Thus
G   is 3-edge colorable (the colors being taken from
$\Gamma - \{e\}$.)

  Now assume that   G   is 3-edge colored, with the elements of $\Gamma - \{e\}$.   Let   R   be any region of   G,   and set
$c(R) = e$.   Let   S   be any other region of   G;   we determine $c(S)$ as follows.   Let   C'   be an arc joining a
point in   R   with a point in   S,   with $C' \cap V(G) = \emptyset$.
Suppose   C'   crosses edges $x_1, x_2, \ldots, x_n$ (repetition

allowed). Set $c(S) = \sum_{i=1}^{n} c(x_i)$. To show that $c(S)$ is
well-defined, we must show that this element of Γ is independent of the particular arc C' selected. Equivalently, we show that if C is a simple closed curve
crossing edges $y_1, y_2, \ldots, y_m$, with $C \cap V(G) = \phi$,
then $\sum_{i=1}^{m} c(y_i) = e$. Let C be such a curve. If $V(G) \cap$
Int $C = \phi$, then each edge crossed by C is crossed an
even number of times; hence $\sum_{i=1}^{m} c(y_i) = e$, in this
case. If $V(G) \cap$ Int $C \neq \phi$, we assume (without loss of
generality) that each edge crossed by C is crossed
exactly once. Let $y_i \subseteq$ Int C, $i = m + 1, \ldots, r$. By
hypothesis, the sum of the colors on three edges incident
with any vertex is e; hence the total of such sums for
the set $V(G) \cap$ Int C is also e; but this sum is also
given by

$$\sum_{i=1}^{m} c(y_i) + 2 \sum_{i=m+1}^{r} c(y_i) = \sum_{i=1}^{m} c(y_i).$$

Therefore $\sum_{i=1}^{m} c(y_i) = e$ in this case also, and $c(S)$
is well-defined.

Now consider two adjacent regions, R_x^1 and R_x^2. We
have assigned colors to the regions of G so that
$c(R_x^2) - c(R_x^1) = c(x) \neq e$; i.e. $c(R_x^2) \neq c(R_x^1)$. Thus G
is 4-region colored.

We will not discuss the four color conjecture further here,
but will move on to the five-color theorem.

8-3. The Five-Color Theorem

The proof below is found in [4].

<u>Thm. 8-7</u>. Five colors will suffice to color any map on the sphere;
i.e. $\chi(S_0) \leq 5$.

Proof: We use induction on p, the order of the graph
G, to show that if $\gamma(G) = 0$, then $\chi(G) \leq 5$. The
anchor at p = 1 is obvious. Now assume that all planar
graphs with p-1 vertices (p > 1) are 5-colorable.
Let G be planar, with p vertices. By Lemma 5-19, G
contains a vertex v of degree 5 or less. By the induction hypothesis, $\chi(G-v) \leq 5$; denote the colors in a
5-coloring of G-v by 1,2,3,4,5. If not all five
colors are used for the vertices adjacent to v in G,
we can color v with one of the colors not so used, to
give $\chi(G) \leq 5$. Otherwise, d(v) = 5, and all five
colors are used for vertices adjacent to v. We can assume that the situation around v is as in Figure 8-3,
and that v_i is colored with color i. Consider now

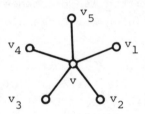

Figure 8-3.

any two colors assigned to non-consecutive vertices v_i,
say 1 and 3, and let H be the subgraph of G - v
induced by all those vertices colored 1 or 3. If
v_1 and v_3 belong to different components of H, then
by interchanging the colors in the component of H containing v_1, say, a 5-coloring of G - v is produced
in which no vertex adjacent with v is assigned the
color 1, and we can use 1 for v. If, on the other
hand, v_1 and v_3 are joined by a path in H, the
above argument guarantees that we can recolor v_2 with
4, and use 2 for v. This completes the proof.

8-4. Other Map-coloring Problems;
The Heawood Map-coloring Theorem

Now let us consider other subspaces of E^3 in which to pose map-coloring questions such as that above, for the sphere (and plane). Strangely enough, if we allow 3-dimensional countries, arbitrarily many colors may be needed to color the map. This is indicated by Figure 8-4, in which the countries are numbered; it is seen that each country meets each of the other countries. (In general, n rectangular parallelopipeds are laid across n other such solids.)

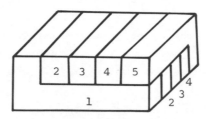

Figure 8-4

Perhaps it seems natural, since the coloring problem is apparently extraordinarily difficult for the sphere, and admits no finite answer in E^3, to consider next the surfaces S_k as candidates for maps and the corresponding map-coloring questions.

The Heawood Map Coloring Theorem (formerly the Heawood Map-Coloring Conjecture) has a particularly colorful background, as outlined in Chapter 1; also see J. W. T. Youngs [74]. We state the theorem first for the orientable case:

Thm. 8-8.

$$\chi(S_k) = \left\lceil \frac{7 + \sqrt{1 + 48k}}{2} \right\rceil, \quad \text{for } k > 0.$$

Note what happens if we replace k with 0 in this formula. This leads some mathematicians to feel that the four color conjec-

ture is probably true. We will give an "equally convincing" argument before the end of this chapter (following Theorem 8-15) to show
that it is probably false!

The corresponding map-coloring question can also be asked for
the closed non-orientable surfaces $\tilde{S}_k$ (spheres with k cross-
caps). Ringel [52] showed the following (the case k = 2 was
solved by Franklin [22]):

<u>Thm. 8-9</u>. $\chi(\tilde{S}_k) = \left\lceil \dfrac{7 + \sqrt{1 + 24k}}{2} \right\rceil$,   for   k = 1   and   k $\geq$ 3;
$\chi(\tilde{S}_2) = 6$.

For example, the formula gives $\chi(\tilde{S}_1) = 6$ (for the projective
plane). Figure 8-5 shows K_6 imbedded in $\tilde{S}_1$, indicating that
$\chi(\tilde{S}_1) \geq \chi(K_6) = 6$.

Recalling that the euler characteristics for S_k and $\tilde{S}_k$ are
given by n = 2 - 2k and n = 2 - k respectively, we can combine
Theorems 8-8 and 8-9 as follows:

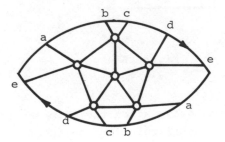

Figure 8-5

<u>Thm. 8-10</u>.   Let M_n be a closed 2-manifold, other than the sphere
or klein bottle, of characteristic n;   then

$$\chi(M_n) = \left\lceil \frac{7 + \sqrt{49 - 24n}}{2} \right\rceil.$$

For ease of notation, we let $f(n) = \dfrac{7 + \sqrt{49 - 24n}}{2}$.   We will
now establish what Heawood knew in 1890:

$$\chi(M_n) \leq [f(n)].$$

We proceed by a series of steps, following the translator's notes in Fréchet and Fan [23].

Lemma 8-11. Let a graph G, with $p \geq 3$, be 2-cell imbedded in M_n, with a denoting the average degree of the vertices of G. Then $a \leq 6(1 - n/p)$.

Proof: Note that $a = 2q/p$. Now $3r \leq 2q = ap$. Also, $p - q + r = n$. Hence

$$q \leq 3(q - r) = 3(p - n),$$

so that

$$a = 2q/p \leq 6(1 - n/p).$$

Thm. 8-12. $\chi(\widetilde{S}_1) \leq 6$.

Proof: We use induction on p, the order of a graph imbedded in $\widetilde{S}_1$. We need only consider graphs having 2-cell imbeddings in $\widetilde{S}_1$, for otherwise (see Youngs [77]), $\gamma(G) = 0$, and $\chi(G) \leq 5$. The result is clearly true for $p \leq 6$. Assume $\chi(G) \leq 6$ for all graphs in $\widetilde{S}_1$ with $p - 1$ vertices, $p \geq 7$; let G be a graph imbedded in $\widetilde{S}_1$, with p vertices. By Lemma 8-11, $a < 6$, so that G has a vertex v such that $d(v) \leq 5$. Then $G - v$ is imbedded in $\widetilde{S}_1$, and $\chi(G - v) \leq 6$, by the induction hypothesis. Since there are vertices of at most five colors adjacent to v, the sixth color can be used for v, and $\chi(G) \leq 6$.

Note that Theorem 8-12, together with Figure 8-5, show that $\chi(\widetilde{S}_1) = 6$.

Thm. 8-13. $\chi(M_n) \leq [f(n)]$, for $n \neq 2$.

Proof: Thanks to Theorem 8-12, we may assume that $n \leq 0$. We use induction on p, to show that $\chi(G) \leq [f(n)]$, if G is imbedded in M_n. (We may assume G to be connected, as the chromatic number of a graph is the largest chromatic number for its components.) It is clear that $\chi(G) \leq [f(n)]$ if $p \leq [f(n)]$. Now assume

that $\chi(G) \leq [f(n)]$ for all graphs with fewer than p
vertices and imbeddable in M_{n_2}. Now, from the definition of $f(n)$, we see that $f^2(n) - 7f(n) + 6n = 0$;
i.e. $6(1 - n/f(n)) = f(n) - 1$. If the imbedding of G
in M_n is 2-cell, then Lemma 8-11 applies, and

$$a \leq 6(1-n/p)$$

$$\leq 6(1-n/f(n))$$

$$= f(n) - 1.$$

If the imbedding is not 2-cell, then it is not minimal
(see again Youngs [77]), and we can find a 2-cell imbedding in M_m, where $m > n$. We then apply Lemma
8-11 as above, to get $a \leq f(m) - 1 \leq f(n) - 1$. Thus
in either case $a \leq f(n) - 1$, and we can find a vertex
v of G having $d(v) \leq [f(n)] - 1$, so that, (using
$\chi(G-v) \leq [f(n)]$), $\chi(G) \leq [f(n)]$. This completes the
proof.

The task remains to show that $\chi(M_n) \geq [f(n)]$, for $M_n \neq \tilde{S}_2$,
the klein bottle. This is done by finding a graph G imbeddable in
M_n and having $\chi(G) = [f(n)]$. For $M_2 = S_0$, K_4 is such a graph;
for $M_1 = \tilde{S}_1$, take $G = K_6$ (as in Figure 8-5); for $M_0 = S_1$, pick
$G = K_7$ (see Figure 8-6 for the dual of K_7 in S_1); for $M_{-2} = S_2$,
let $G = K_8$. In fact, for $M_n \neq \tilde{S}_2$, $[f(n)]$ is attained by the

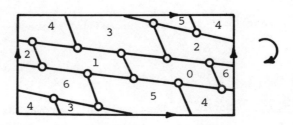

Figure 8-6

largest complete graph imbeddable in M_n. We now confine our attention to the orientable case and explore this claim in some detail.

Let us assume the truth of the complete graph theorem (which
will be discussed in detail in Chapter 9):

$$\gamma(K_m) = \left\{\frac{(m-3)(m-4)}{12}\right\}, \quad \text{for} \quad m \geq 3.$$

From this it will follow that $\chi(M_n) \geq [f(n)]$ (in the orientable
case; the non-orientable case is handled similarly).

<u>Thm. 8-14.</u> $\chi(S_k) \geq [f(2-2k)] = \left[\frac{7 + \sqrt{1 + 48k}}{2}\right]$.

Proof: Consider S_k. Define $m = [f(2-2k)]$, and now
consider also $S_{\gamma(K_m)}$. Note that $\gamma(K_m) \leq k$, so that
$\chi(S_{\gamma(K_m)}) \leq \chi(S_k)$. Now K_m imbeds in $S_{\gamma(K_m)}$. Clearly
$\chi(S_{\gamma(K_m)}) \geq m = [f(2-2k)]$, so that $\chi(S_k) \geq [f(2-2k)]$.

Theorems 8-13 and 8-14 combine to prove Theorem 8-8, with the
understanding that it remains to establish the formula for the
genus of K_m. Before indicating how this is done (in the next
chapter), we pause for two related results.

8-5. A Related Problem

We have seen that, for any closed 2-manifold except the sphere,
the maximum chromatic number of an imbedded graph is taken on by a
complete graph. We now show that the complete graphs play the same
role with respect to maximizing the minimum degree of an imbedded
graph.

<u>Thm. 8-15.</u> Let M_n be a closed 2-manifold of characteristic n,
and G a graph. If G has an imbedding in M_n, then
$\delta(G) \leq g(n)$, where

$$g(n) = \begin{cases} \left[\dfrac{5 + \sqrt{49 - 24n}}{2}\right], & \text{if} \quad n < 2 \\[4mm] 5, & \text{if} \quad n = 2. \end{cases}$$

Furthermore, there exists a graph G,   imbeddable in
M_n ,   such that $\delta(G) = g(n)$.

Proof: The theorem is known to be true for   n = 2,   as
Lemma 5-18 and the icosahedral graph show.   Suppose now
that   G   is a graph having   p   vertices and   q   edges,
with $\delta(G) > g(n)$,   and a 2-cell imbedding in $M_n (\neq S_0)$.
By standard arguments, $2q \geq 3r$,   and also $2q \geq$
$p(g(n) + 1)$.   We may assume that   G   is connected, since
if the theorem is true for every component of   G,   it is
also true for   G.   The euler formula applies, so that

$$n = p - q + r$$

$$\leq \left(\frac{2}{g(n)+1} - \frac{1}{3}\right)q$$

$$= \left(\frac{5 - g(n)}{3(g(n)+1)}\right)q.$$

We may assume that $n \leq 0$,   as the above inequality is
clearly impossible for   n = 1.   But for $n \leq 0$, $g(n) \geq 6$,
so that

$$q \leq \frac{-3n(g(n)+1)}{g(n) - 5} .$$

But since $\delta(G) > g(n)$, $p \geq g(n) + 2$,   and

$$2q \geq (g(n)+2)(g(n)+1) .$$

We note that

$$(g(n)+2)(g(n)-5) = \left[\frac{9 + \sqrt{49-24n}}{2}\right]\left[\frac{-5 + \sqrt{49-24n}}{2}\right]$$

$$> \left(\frac{7 + \sqrt{49-24n}}{2}\right)\left(\frac{-7 + \sqrt{49-24n}}{2}\right)$$

$$= -6n.$$

It now follows that

$$q \geq \frac{(g(n)+2)(g(n)+1)}{2}$$

$$> \frac{-3n(g(n)+1)}{g(n) - 5}$$

$$\geq q,$$

a contradiction. Hence $\delta(G) \leq g(n)$.

Now suppose that G has a non 2-cell imbedding in
M_n. By a result of Youngs [77], G has a 2-cell im-
in some $M_{n'}$, where $n < n'$. From what we have shown
above, $\delta(G) \leq g(n') \leq g(n)$.

Ringel and Youngs have shown [56] (also, see
Chapter 9) that the complete graph $K_{[g(2-2k)]+1}$ is
imbeddable in S_k, for $k \geq 1$. Ringel [27] has shown
that the complete graph $K_{[g(2-k)]+1}$ is imbeddable in
$\tilde{S}_k$, for all positive k except $k = 2$. It remains
to find a graph G imbeddable in $\tilde{S}_2$ and having
$\delta(G) = 6$. We begin by considering two projective planes,
P_1 and P_2, each with a complete graph K_5 imbeded
as indicated in Figure 8-7. Cut open disks D_1 and D_2
from the interiors of the five-sided regions of P_1 and
P_2, respectively. Let T be a cylinder disjoint from
P_1 and P_2, each with a complete graph K_5 imbedded
and C_2. Identify C_1 with the boundary of D_1 and
C_2 with the boundary of D_2. The result, $(P_1-D_1) \cup$
$T \cup (P_2-D_2)$, is a klein bottle (see Problem 8-7). The
graph G is then constructed by adding the edges
(i,i'), $(i,i'+1)$, $i = 1,2,3,4,5$ (where the vertex 6'
is the same as the vertex 1'). This completes the
proof.

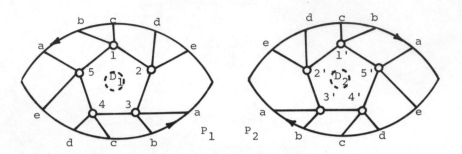

Figure 8-7

We thus make the following observation. The sphere is the only
closed orientable 2-manifold for which the maximum minimum degree
is not attained by a complete graph. We have also seen that for
every closed 2-manifold (whether orientable or non-orientable) except the sphere, the maximum chromatic number is attained by a
complete graph. Should the sphere be shown to fail in this regard,
the four color conjecture would be disaffirmed.

8-6. A Four-Color Theorem for the Torus

Thus far in this chapter we have been discussing, for a given
closed 2-manifold M, the chromatic number of arbitrary graphs that
can be imbedded in M. In this section we impose a restriction on
the girth of the graphs we are considering.

Def. 8-16.   The girth $g(G)$ of a graph G is the length of a
             shortest cycle (if any) in G.

Thus a graph G with cycles but no triangles has $g(G) \geq 4$;
if G is a forest, we write $g(G) = \infty$. The following theorem was
shown by Grötzsch [25]:

Thm. 8-17.   If $\gamma(G) = 0$ and $g(G) \geq 4$, then $\chi(G) \leq 3$.

The graph $G = C_5$ shows that equality can hold in Theorem 8-17. In this section we [35] find an upper bound for the chromatic number of toroidal graphs having no triangles, and show that this bound is best possible. We also consider toroidal graphs of arbitrary girth.

Def. 8-18. A connected graph G is said to be <u>n-edge-critical</u> $(n \geq 2)$ if $\chi(G) = n$ but, for any edge x of G, $\chi(G-x) = n - 1$.

The next theorem is due to Dirac [17].

Thm. 8-19. If G is n-edge-critical, $n \geq 4$, and if $G \neq K_n$, then $2q \geq (n-1)p + n - 3$.

We are now able to find the analogue of Grötzsch's Theorem, for the torus.

Thm. 8-20. If $\gamma(G) \leq 1$ and $g(G) \geq 4$, then $\chi(G) \leq 4$.

Proof: Let $\chi(G) = n \geq 5$. We first assume that G is n-edge-critical, and hence connected. Since $g(G) \geq 4$, $G \neq K_n$. By Theorem 8-19,

$$2q \geq (n-1)p + n-3.$$

Now if $\gamma(G) = 1$, then by Corollary 6-15,

$$4p \geq 2q \geq (n-1)p + n-3;$$

thus $n \leq 4$. If $\gamma(G) = 0$, then $n \leq 3$, by Theorem 8-17. In either case we have a contradiction, so that $n \leq 4$.

Now suppose that G is not n-edge-critical. Then G contains an n-edge-critical subgraph H, and the argument above shows that $\chi(G) = \chi(H) = n \leq 4$.

The graph of Figure 8-8, constructed by Mycielsky [44] as an example of a graph having no triangles and chromatic number four, also has genus one, so that the bound of Theorem 8-20 cannot be improved.

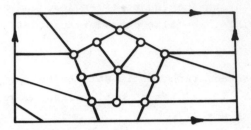

Figure 8-8

The situation for the torus is (almost) completely analyzed in:

<u>Thm. 8-21</u>. If $\gamma(G) \leq 1$ and $g(G) = m$, then

$$\chi(G) \leq \begin{cases} 7, & \text{if } m = 3 \\ 4, & \text{if } m = 4 \text{ or } 5 \\ 3, & \text{if } m \geq 6. \end{cases}$$

Moreover, all the bounds are sharp, except possibly for m = 5.

<u>Proof</u>: If $m \geq 6$, then each region in an imbedding for G has at least six edges in its boundary, so that $2q \geq 6r$. As in the proof of Theorem 8-20, we may assume that $\gamma(G) = 1$ and that G is n-edge-critical, where $n = \chi(G)$. If $n \geq 4$, then $2q \geq 3p + 1$, by Theorem 8-19. Then, by Corollary 5-14,

$$0 = p - q + r$$

$$\leq \frac{2q-1}{3} - q + \frac{q}{3}$$

$$= -\frac{1}{3},$$

an obvious contradiction. Hence for $\gamma(G) \leq 1$ and $g(G) \geq 6$, we must have $\chi(G) \leq 3$. This bound is best

possible, as an appropriate subdivision G of the
Petersen graph (shown imbedded in S_1 in Figure 8-9)
can always be found, having $g(G) = m(m \geq 5)$, $-\gamma(G) = 1$,
and $\chi(G) = 3$.

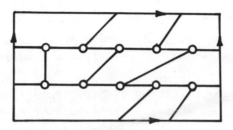

Figure 8-9

For $m = 4$ or 5, it follows from Theorem 8-19 that
$\chi(G) \leq 4$.   (Now, see Problem 8-9.)   Figure 8-8 shows
that equality can hold for $m = 4$.   For $m = 3$, we
refer to the Heawood Map-Coloring Theorem.

8-7. k-degenerate Graphs

        Before getting to the focal point of this text,   in the next
(and last) chapter, we digress briefly.   The generalization below
of Theorem 8-10 may be of interest.
        A coloring number for graphs closely related to the chromatic
number is the vertex-arboricity (see [14].)

Def. 8-22.   The <u>vertex</u> <u>arboricity</u>, $a(G)$,   <u>of a graph</u> G is the
            minimum number of subsets that $V(G)$ can be partitioned
            into so that each subset induces an acyclic graph.

Def. 8-23.   The <u>vertex</u> <u>arboricity</u> <u>of a surface</u> S_k is the maximum
            vertex-arboricity among all graphs which can be imbed-
            ded in S_k.

In 1969, Kronk [34] showed that the vertex arboricity of S_k,
$k > 0$, is $\left\lceil \dfrac{9 + \sqrt{1 + 48k}}{2} \right\rceil$. Chartrand and Kronk [13], also in 1969,
proved that the vertex-arboricity of the sphere is three.   The
similarity of Kronk's result to those of Ringel and of Ringel and
Youngs for the chromatic number suggested the generalization mentioned below.

Def. 8-24.   A graph G is said to be k-degenerate if every induced
             subgraph H of G satisfies the inequality $\delta(H) \leq k$.

Def. 8-25.   The vertex partition number, $\rho_k(G)$, of a graph G is
             the minimum number of subsets into which V(G) can be
             partitioned so that each subset induces a k-degenerate
             subgraph of G.

The parameters $\rho_0(G)$ and $\rho_1(G)$ are the chromatic number and
vertex arboricity of G, respectively (see Problem 8-4).   A general
study of k-degenerate graphs has been begun in [38], where many of
the well-known results for the chromatic number and the vertex-
arboricity of a graph have been extended to the parameters $\rho_k(G)$,
for all non-negative integers k.

Def. 8-26.   The vertex partition number of the closed 2-manifold M_n,
             denoted by $\rho_k(M_n)$, is the maximum vertex partition
             number $\rho_k(G)$ of all graphs G which can be imbedded
             in M_n.

The following theorem (for a complete proof, see [39]) almost
completely generalizes the results of Kronk, Ringel, and Ringel and
Youngs mentioned above.

Thm. 8-27.   The vertex partition numbers for a closed 2-manifold M_n
             are given by the formula:

$$\rho_k(M_n) = \left\lceil \frac{(2k+7) + \sqrt{49-24n}}{2k + 2} \right\rceil,$$

             where k = 0,1,2,...; and n = 2,1,0,-1,-2,..., except
             for the following cases:

(i)   in the orientable case, $\rho_0(S_0) = 4$ or 5,
$\rho_1(S_0) = 3$, $\rho_3(S_0) = \rho_4(S_0) = 2$; and

(ii) in the non-orientable case, $\rho_0(\tilde{S}_2) = 6$, $\rho_1(\tilde{S}_2) = 3$
or 4, $\rho_2(\tilde{S}_2) = 2$ or 3.

We make the following comments about the proof of Theorem 8-27.
Set $f(k,n) = \left[\dfrac{(2k+7) + \sqrt{49-24n}}{2k + 2}\right]$. The proof is divided into three
parts:

(i) $\rho_k(M_n) \le f(k,n)$,   for $M_n \ne S_0$ (the proof breaks
down for the sphere).

(ii) $\rho_k(M_n) \ge f(k,n)$,   for $M_n \ne \tilde{S}_2$ (the proof fails
for the klein bottle).

(iii) the exceptional cases are treated separately:

(a)   for $M_n = S_0$, $\rho_0(S_0) = 4$ or 5, $\rho_1(S_0) = 3$
appear in the literature; for $\rho_k(S_0) = 2$,
$k = 2,3,4$,   see Problem 8-5; finally, $\rho_k(S_0) = 1$
for $k \ge 5$,   since any planar graph is 5-
degenerate, by Lemma 5-18.

(b)   For $M_n = \tilde{S}_2$,   additional ad hoc arguments are
devised. For example, the graph constructed
in Figure 8-7 shows that $\rho_5(\tilde{S}_2) \ge 2$;   from
(i) we see that $\rho_5 = (\tilde{S}_2) \le 2$;   thus
$\rho_5(S_2) = 2$.

8-8.   Coloring Graphs on Pseudosurfaces

The pseudosurfaces $S(k; n_1(m_1),\ldots,n_t(m_t))$ have been defined
in Section 5-5 and re-encountered in Sections 6-7 and 6-8. Dewdney
[16] has studied a subclass of these pseudosurfaces, namely those of
the form $S(0,n(2))$:

Def. 8-28.   The <u>chromatic number</u>, $\chi(S(0;n(2)))$, <u>of the pseudo-</u>
<u>surface</u> $S(0;n(2))$ is the largest chromatic number
$\chi(G)$ of any graph G that can be imbedded in $S(0;n(2))$.

Thm. 8-29. $\chi(S(0;n(2))) \leq n + 4$,   for $n > 0$;   equality holds for
             $n = 1, 2, 3, 4.$*

For example, Figure 6-6 shows K_5 imbedded in $S(0;1(2))$,
showing that $\chi(S(0;1(2)) \geq 5$. Similarly, K_6 imbeds in
$S(0;2(2))$, to give equality for the case $n = 2$. Note that we
state this coloring problem for graphs rather than for maps; the
dual of G in $S(0;n(2))$ is not a 2-cell imbedding, so that there
is not the natural correspondence we find for surfaces. (*The cases
$n = 3$ and 4 were established by Mark O'Bryan and James Williamson
respectively.)

8-9. Problems

8-1.)   Let $G \neq \overline{K_n}$ be a bipartite graph. Show that $\chi(G) = 2$.
8-2.)   Find $\chi(C_n)$, for any cycle C_n.
8-3.)   Find an imbedding of K_7 on S_1. Form the dual of this
        imbedding, and explain why this shows that $\chi(S_1) \geq 7$.
8-4.)   Show that $\rho_0(G) = \chi(G)$ and that $\rho_1(G) = a(G)$.
8-5.)   Show that $\rho_k(S_0) = 2$, for $k = 2,3,4$. (Hint: for each k,
        use induction to show that $\rho_k(S_0) \leq 2$. Then consider graphs
        of certain regular polyhedra.)
8-6.)* Show (as Ringel did) that any map on the surface of the
        sphere, in which each country has at most two components, can
        be colored with 12 colors. (Hint: it may be helpful to show
        that if a graph G is n-critical , then $\delta(G) \geq n - 1$; i.e.
        if $\chi(G) = n$, but $\chi(G-v) = n - 1$, for all vertices v in
        G. Then form two "dual" graphs for an arbitrary map, one
        where the vertices represent countries, the other with ver-
        tices representing regions of land. Use also the fact that
        $q \leq 3p - 6$, for planar connected graphs.)
8-7.)   Show that the connected sum of two projective planes (as in
        the proof of Theorem 8-15) is a klein bottle. (Hint: find
        the characteristic of the resulting closed 2-manifold, using
        the graph G constructed in the same proof.)
8-8.)   Show that $\chi(G) = 4$, for the graph G of Figure 8-8.
8-9.)** Does there exist a toroidal graph G having $g(G) = 5$ and
        $\chi(G) = 4$?
8-10.)** Prove or disprove: K_9 imbeds in $S(0;5(2))$ (and hence
        $\chi(S(0;5(2))) = 9$.) Is $\chi(S(0;n(2))) = n + 4$ for all n?
        Compare Problem 6-7.

8-11.) Devise a reasonable definition of $\chi(\Gamma)$, the "chromatic number of a group", and obtain some elementary results about this parameter. Extend your definition to $\rho_k(\Gamma)$, for arbitrary vertex partition numbers.

Chapter 9: Quotient Graphs and Quotient Manifolds
(and Quotient Groups!)

In this chapter we present the beautiful theory of quotient graphs and quotient manifolds. This theory was introduced by Gustin [26], developed by Youngs (see, for example, [74], [75], and [78]), and used by Ringel and Youngs to find the genus of K_n and thus solve the Heawood map-coloring problem, as explained in the previous chapter. The application of the theory to the graphs K_n falls into 12 cases, depending upon the residue modulo 12 of n. The theory applies directly, for $n \equiv 0, 3, 4, 7$ (mod 12), as will be seen shortly. For the remaining eight cases, the theory is modified (by the theory of vortices) to complete the solution. We will treat the case $n \equiv 7$ (mod 12) completely, and discuss the case $n \equiv 10$ (mod 12); this will give an indication of the power and beauty of the theory. The remaining ten cases are treated similarly, although many complicating details must be handled properly. (Perhaps one should expect a complicated solution, to a complicated problem!)

We will then see how the theory (designed to produce triangular imbeddings for K_n) may be extended to handle first triangular imbeddings for Cayley graphs in general, and then to handle regular imbeddings $(r = r_n, n \geq 3)$ in general, for Cayley graphs. This is the scope of the theory, as announced by Gustin. But Youngs' theory of vortices [75] hints at an even more general theory; we present this general theory, as unified by Jacques [32]. We conclude the text with a sampling of applications of this powerful theory.

9-1. The Genus of K_n

Let us now turn our attention to the complete graphs K_n. Recall that if we can show that

$$\gamma(K_n) = \left\{ \frac{(n-3)(n-4)}{12} \right\}, \quad n \geq 3,$$

the Heawood map-coloring theorem will be established. We see the origin of the number on the right-hand side of the above equality in the following:

<u>Thm. 9-1.</u> Let K_n be minimally imbedded in a surface M. Then

$$\gamma(K_n) = \gamma(M) = \frac{(n-3)(n-4)}{12} + \frac{1}{6} \sum_{i \geq 4} (i-3) r_i.$$

<u>Proof</u>: From Corollary 6-14, we know that

$$\gamma(K_n) \geq \frac{n(n-1)}{12} - \frac{n}{2} + 1 = \frac{(n-3)(n-4)}{12},$$

with equality if and only if K_n has a triangular imbedding. But we can be more specific than this; we can get some information about the non-triangular regions (if any!). If K_n is minimally imbedded in M, then clearly $\gamma(K_n) = \gamma(M)$, and the imbedding is 2-cell, by Theorem 6-11. Thus the euler formula applies, and

$$\gamma(M) = 1 - 1/2(p - q + r)$$

$$= 1 - 1/2\left(n - 1/2 \sum_{i \geq 3} i r_i + \sum_{i \geq 3} r_i\right)$$

$$= 1 - \frac{n}{2} + 1/4 \sum_{i \geq 3} (i-2) r_i$$

$$= 1 - \frac{n}{2} + 1/4 \sum_{i \geq 3} r_i + 1/4 \sum_{i \geq 4} (i-3) r_i$$

$$= 1 - \frac{n}{2} + \frac{r}{4} + 1/4 \sum_{i \geq 4} (i-3) r_i$$

$$= 1 - \frac{n}{2} + 1/4 (2 - 2\gamma(M) + \frac{n(n-1)}{2} - n) + 1/4 \sum_{i \geq 4} (i-3) r_i;$$

solving this equation for $\gamma(M)$, we get the result.

We now see that if K_n has a triangular imbedding $(r_i = 0,$
$i \geq 4)$, then

$$\gamma(K_n) = \frac{(n-3)(n-4)}{12} ,$$

and $(n-3)(n-4) \equiv 0 \pmod{12}$; i.e. $n \equiv 0, 3, 4, 7 \pmod{12}$. Moreover, in general, $\gamma(K_n) = \left\{ \frac{(n-3)(n-4)}{12} \right\}$, if we can show that
$\sum_{i \geq 4} (i-3)r_i \leq 5$. It is now perhaps apparent why there are twelve
cases for the determination of $\gamma(K_n)$, and why only four of them
admit triangular imbeddings. Let us consider these four cases now.

What is needed is a method of constructing triangular imbeddings.
The naive trial-and-error method easily handles $n = 3, 4,$ and 7;
it becomes a bit sticky at $n = 12$. We turn away from the drawing
board and employ the algebraic description of 2-cell imbeddings
given us by Edmonds' permutation technique. Now we seek a means of
selecting judiciously the local vertex permutations $(p_1, p_2, \ldots, p_n)$;
this is what the method of quotient graphs and quotient manifolds is
all about!

9-2. The Theory of Quotient Graphs and
and Quotient Manifolds, as Applied to K_n

We introduce this theory by means of the example K_7. Let K_7
be imbedded in S_1; by Theorem 9-1, this imbedding must be a tri-
angulation. Select a group Γ for which K_7 is a Cayley color
graph; in this case, we can only pick $\Gamma = Z_7 = \langle x \mid x^7 = e \rangle$, but we
pick x, x^2, x^3 as generators for Γ. Label the vertices of K_7 with
the elements of Γ (one should also think of the edges as being
directed and colored appropriately.) Now take the dual of this imbedding; assume this is as pictured in Figure 8-6. Each region in
the dual (formerly a vertex of K_7) is now labeled with a distinct
group element: 0, 1, 2, 3, 4, 5, or 6. We proceed to label the
boundary edges of each region of the dual, as indicated in Figure
9-1. (Note that $(g^{-1}h)^{-1} = h^{-1}g$.) We observe that the seven
regions of the dual have identical boundaries: 1, 3, 2, 6, 4, 5.

Figure 9-1

We summarize this information in a map having one region, as shown in Figure 9-2. But $1^{-1} = 6$, $2^{-1} = 5$, and $3^{-1} = 4$; thus the six

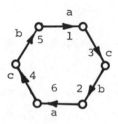

Figure 9-2

edges have a natural identification, in three pairs; we make this identification, to form a closed orientable 2-manifold, as in Figure 9-3. The result (in this case, S_1) is the _quotient_ _manifold_; the corresponding graph (actually, in this case, it is a pseudograph) is the _quotient_ _graph_. The subgroup of Γ consisting of all vertices

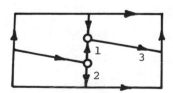

Figure 9-3

of K_7 whose regions in the dual had the same ordering of directed
edges in their boundaries as did $e = 0$ (in this case, Z_7 itself)
gives rise to the <u>quotient group</u> (in this case, the trivial group).
The index of this subgroup in Γ (in this case, 1) is the <u>index
of the imbedding</u>. The point is this: <u>all the information needed to
describe a triangular imbedding of</u> K_7 <u>in</u> S_1 <u>is contained in the
quotient graph</u>, imbedded in its quotient manifold (which, after all,
was obtained by "modding out" the subgroup Z_7).

To see this, let the permutation at vertex 0 be given by the
boundary of the single region in the quotient manifold:

$$0: 1, 3, 2, 6, 4, 5.$$

The remaining local vertex permutations may be obtained by succes-
sively adding 1 to every entry in this row (remember, we are in
the group Z_7!):

$$
\begin{aligned}
0: &\quad 1, 3, 2, 6, 4, 5 \\
1: &\quad 2, 4, 3, 0, 5, 6 \\
2: &\quad 3, 5, 4, 1, 6, 0 \\
3: &\quad 4, 6, 5, 2, 0, 1 \\
4: &\quad 5, 0, 6, 3, 1, 2 \\
5: &\quad 6, 1, 0, 4, 2, 3 \\
6: &\quad 0, 2, 1, 5, 3, 4.
\end{aligned}
$$

Now, compute orbits (corresponding to regions in a 2-cell imbedding
of K_7):

0-1-5	1-2-6	3-5-6
0-2-3	1-3-4	3-6-4
0-3-1	1-4-2	
0-4-6	1-6-5	
0-5-4	2-4-5	
0-6-2	2-5-3.	

We see that we have an imbedding of K_7 for which $r = r_3 = 14$;
that is -- a triangular imbedding. This is no accident; it is

guaranteed by the theory of quotient graphs and quotient manifolds!

    We now examine this theory more closely. (At first we consider only triangular imbeddings for   K$_n$;   in sections 9-4 - 9-7, we generalize.)   Recall that, for a graph   K, D* = {(u,v)|[u,v] ∈ E(K)}; in this chapter, we set   D* = K*.

Def. 9-2.   A <u>current graph</u> is a triple   (K,Γ,λ),   where   K   is a
           pseudograph,   Γ   is a finite group with identity   e,   and
           λ: K* → Γ - e   is a map such that $(\lambda(a))^{-1} = \lambda(a^{-1})$,
           for all   a ∈ K*.

Note that a Cayley color graph is a current graph.

Def. 9-3.   Let   K   be 2-cell imbedded in a surface   M,   with
           v ∈ V(K).   We say that <u>Kirchoff's Current Law</u> (KCL) holds
           at   v   if the product of the currents directed away from
           v,   taken in the order given by   p$_v$,   is the identity,
           e.

Def. 9-4.   Given a group   Γ   of order   n   and a subgroup   Ω   of
           order   m,   a <u>quotient manifold</u>   M(Γ/Ω)   is a closed 2-
           manifold having the following properties:

        1.)   M(Γ/Ω)   is oriented and has a 2-cell decomposition,
             given by a pseudograph   K.
        2.)   The 2-cells of the decomposition are   N = n/m   in
             number, are named   [X], X ∈ Γ/Ω,   and each is an
             (n-1)-gon.   N   is called the <u>index</u> of   M(Γ/Ω).
        3.)   The oriented edges in   K   carry currents from   Γ - e.
        4.)   K   is a current graph.
        5.)   For   X ∈ Γ/Ω,   each element of   Γ - e   appears
             exactly once as a current on some edge of |X|*,
             the oriented boundary of   [X]   (with orientation in-
             herited from   M(Γ/Ω).)
        6.)   [X]   meets   [Y] ≠ [X]   along those edges of   [X]*
             whose currents are in the set $X^{-1}Y$,   and nowhere
             else.
        7.)   [X]   meets itself along those edges of   [X]*   whose
             currents are in the set $(X^{-1}X)$ - e,   and nowhere
             else.   Moreover, the meeting is along a <u>singular</u>
             <u>edge</u> (a loop), if and only if the current is of
             order   2.

8.)   Each vertex of K is of degree 1 or 3.

9.)   The current going into each vertex of degree 1 is
      of order 3.

10.) The KCL holds at each vertex of degree 3.

<u>Thm. 9-5</u>. If K_n is triangularly imbedded in a surface M, then
         for any group Γ of order n and 1-1, onto map
         A: $V(K_n) \rightarrow \Gamma$, there exists a subgroup Ω and a quotient
         manifold $M(\Gamma/\Omega)$.

It is clear that $\Omega = \{e\}$ will always satisfy Theorem 9-5;
but this gives no progress for the imbedding problem, since
$M(\Gamma/\{e\}) = M$. Economy of effort here is inversely proportional to
the index of the imbedding.

The graph K of Definition 9-4 is called a <u>quotient</u> <u>graph</u>; it
is denoted also by $S(\Gamma/\Omega)$. Note that $S(\Gamma/\Omega)$ is not the graph
alone, but the graph together with a collection of local vertex
permutations that gives rise to an imbedding of the graph in a
quotient manifold.

For additional examples of quotient graphs and quotient manifolds, refer to [74] or to subsequent sections of this chapter.

9-3.   The Genus of K_n (again)

It is the next theorem which is basic to determining $\gamma(K_n)$.

<u>Thm. 9-6</u>. If $S(\Gamma/\Omega)$ is a quotient graph, where Γ has order n,
         then there exists a triangular imbedding of K_n.

Thus if we can find an appropriate quotient graph (and this may
be a considerably easier task, if Ω is of low enough index), we
have found a combinatorial equivalent of a triangular imbedding.

The justification for the above theory is given very nicely by
Youngs. We prove here only the special case of an index one imbedding, with $\Gamma = Z_n$.

<u>Thm. 9-7</u>. If $K = S(Z_n/Z_n)$ is a quotient graph such that

         (i) K is cubic

(ii) $\lambda: K^* \to Z_n - 0$ is 1-1 and onto,

then K_n has a triangular imbedding.

<u>Proof</u>: Since $\Omega = \Gamma$, K is of index one. Hence there
is exactly one region in the cellular decomposition of
the quotient manifold $M(Z_n/Z_n)$. This region must contain every edge $\alpha \in K^*$ in its boundary. Since λ is
onto, for every $g \in Z_n - 0$, there is an edge $\alpha \in K^*$
such that $\lambda(\alpha) = g$. Since λ is 1-1, there is no edge
$\beta \neq \alpha$ such that $\lambda(\beta) = g$. Thus the succession of currents on the boundary of the region is a cyclic permuta-
tion of $Z_n - 0$ of length $(n-1)$; we take this permutation as p_0 (where we label the vertices of K_n by:
$V(K_n) = \{0,1,2,\ldots,n-1\}$.) Suppose p_0 is given by:

$$p_0: (a_1,a_2,\ldots,a_{n-1});$$

then the remaining p_i are given by:

$$p_i: (a_1+i,a_2+i,\ldots,a_{n-1}+i),$$

i = 1, 2, $\ldots$, n-1; all arithmetic is done in $\Gamma = Z_n$.
Note that $a_k+i \neq i$, for otherwise, $a_k = 0$. Hence the
collection $(p_0,p_1,\ldots,p_{n-1})$ determines a 2-cell imbedding of K_n. It remains to show that $r = r_3$ for this
imbedding.

        Now suppose that $p_0(a) = b$, and let $\alpha,\beta \in K^*$,
with $\lambda(\alpha) = a$, $\lambda(\beta) = b$. The situation, by property
(i) of the hypothesis, must be as depicted in Figure 9-4.
(The arrow gives the orientation at the vertices; by
Edmond's algorithm, the region boundaries have the reverse orientation.)

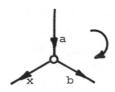

Figure 9-4

But, from the KCL, $a - b - x = 0$, or $x = a - b$.
Now, also, $p_0(-b) = x = a - b$. Therefore, $p_b(0) = x + b = a$. Now let (u,v) be any directed edge in K_n.
We will show that (u,v) is in the boundary of a triangular region, to complete the proof. Compute the orbit, beginning

$$u-v- \quad .$$

Let $p_v(u) = w$; then we have

$$u-v-w- \quad .$$

But since $p_v(u) = w$, $p_0(u-v) = w - v$. Letting $u - v = a$ and $w - v = b$ in the above discussion, we see that

$$p_{w-v}(0) = u - v;$$

hence $p_w(v) = u$, and we have

$$u-v-w-u- \quad .$$

Next, let $v - w = a$ and $u - w = b$; then (since $p_w(v) = u$, $p_0(v-w) = u-w$)

$$p_{u-w}(0) = v - w,$$

and $p_u(w) = v$. Thus

$$u-v-w-u-v$$

constitutes an orbit under P for K_n, of length 3.
This completes the proof.

We next prove that $\gamma(K_n)$ is as predicted, for $n \equiv 7 \pmod{12}$.
By Theorem 9-7, we need only find a cubic quotient graph, $K = S(Z_n/Z_n)$, with a 1-1 onto assignment $\lambda: K^* \to Z_n - 0$. To simplify the presentation of K, instead of showing K imbedded in $M(Z_n/Z_n)$, we draw K in the plane with the following device describing the local vertex permutations: a solid vertex has its incident edges ordered clockwise; a hollow vertex, counterclockwise.

For example, Figures 9-3 and 9-5 depict exactly the same
$K = S(Z_7/Z_7)$.

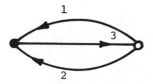

Figure 9-5

<u>Thm. 9-8.</u> K_{12s+7} has a triangular imbedding.

Proof: Let $\Gamma = Z_{12s+7}$. We have already treated the
case $s = 0$. The cases $s = 1$ and $s = 2$ are shown in
Figures 9-6 and 9-7 respectively. The generalization to
all s is as in Figure 9-8, with the vertical edges

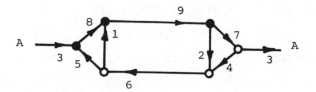

Figure 9-6

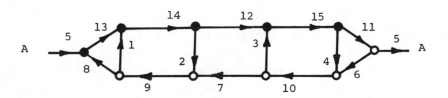

Figure 9-7

directed alternately and carrying the currents 1, 2,
..., 2s consecutively. All other currents are determined by the KCL. It is straightforward to check that K

Figure 9-8

has 6s + 3 edges, or 12s + 6 directed edges, carrying
all the currents from Z_{12s+7} - 0. The single region in
$M(Z_{12s+7}/Z_{12s+7})$ is:

(2s+1) − (5s+3) − (5s+4) −...− (6s+3) − (4s+3) − (2s+2) −

(−2s) − (−6s−3) − (−2s+1) − (2s+3) −...− (−1) − (3s+2) −

(−2s−1) − (−4s−3) − (2s) − (4s+2) − (2s−1) − (−4s−4) −...−

(1) − (−5s−3) − (−3s−2) − (−3s−3) −...− (−4s−2) − (−2s−2).

Thus K is a quotient graph, satisfying the properties
of Theorem 9-7, and K_{12s+7} has a triangular imbedding,
for all non-negative integers s.

Cor. 9-9. $\gamma(K_{12s+7})$ = (3s+1)(4s+1), s ≥ 0.

    The theory of current graphs with vortices is employed to find
$\gamma(K_n)$, for n ≡ 10 (mod 12). The group Z_{12s+7} is used to find a
triangular imbedding for K_{12s+10} - K_3; the current graph in this
case has three vertices of degree one; otherwise it is essentially
the K of Figure 9-8. (See [76].) This is called the regular part
of the problem. Next, the surface in which K_{12s+10} - K_3 is tri-
angularly imbedded is modified -- by the addition of one well-chosen
handle -- so as to accommodate the three edges removed in K_3. This
is called the additional adjacency part of the problem. The final

result is a (non-triangular) imbedding of K_{12s+10} in a surface of
the appropriate genus.

The remaining ten cases for $\gamma(K_n)$ are handled similarly, with
varying degrees of complexity; see [76], [57], [79], [66], [58],
[59], [65], and [43]. A constructive proof is given for each case
but $n \equiv 0 \pmod{12}$; for this case the theory of finite fields sup-
plements the theory of quotient graphs in establishing the <u>existence</u>
of a triangular imbedding (see [65]). We now turn our attention to
graphs other than K_n.

9-4. Extending the Theory

Slight modifications of the above theory allow us to attack
other Cayley color graphs for which triangular imbeddings are possible. In fact, there is also a theory for quadrilateral imbeddings
of Cayley color graphs. Rather than discuss the modifications of
the theory in this section, we will examine three examples; an even
more general theory will be given in the next section.

<u>Example 1</u>: Let us try to find a triangular imbedding for $K_{2,2,2}$.
Since $K_{2,2,2}$ is of order six, there are two possible groups to
work with: Z_6 and S_3. Let's try them both.

Taking $Z_6 = \langle x,y \mid x^6 = y^3 = e \rangle$, we see that $K_{2,2,2}$ is a
Cayley color graph for Z_6. (Here, $x = 1$ and $y = 2$, with $e = 0$;
y is clearly redundant, but its presence as a generator enables us
to picture Z_6 with $K_{2,2,2}$.) We try first for an index one imbedding. This requires a quotient graph K having four directed edges
(corresponding to $1,5,2,4$) and one region in its quotient manifold
But then K has two edges; no such K is possible. Next we try
for an index two imbedding. This requires a quotient manifold
having two quadrilateral regions, and hence a quotient graph with
four undirected, and eight directed, edges. The quotient graph is
shown in Figure 9-9, imbedded in its quotient manifold, the sphere.

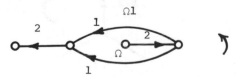

Figure 9-9.

Let $\Omega = \{0,2,4\}$: 1, 5, 4, 2

$\Omega_1 = \{1,3,5\}$: 1, 2, 4, 5.

Then: 0: 1, 5, 4, 2   1: 2, 3, 5, 0

             2: 3, 1, 0, 4   3: 4, 5, 1, 2

             4: 5, 3, 2, 0   5: 0, 1, 3, 4

gives a triangular imbedding of $K_{2,2,2}$.

    The orbits are: 0-1-2   1-3-2

                           0-2-4   1-5-3

                           0-4-5   2-3-4

                           0-5-1   3-5-4.

Figure 9-10 shows this imbedding of $K_{2,2,2}$ (the octahedral graph) in S_0.

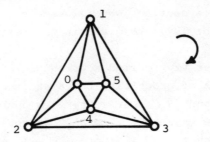

Figure 9-10

Suppose now that we take $\Gamma = S_3 = \langle (23), (12), (132) \rangle$; then $K_{2,2,2}$ is again a Cayley color graph. An index one imbedding _is possible_, in this case, as shown by Figure 9-11, where the

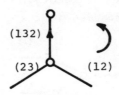

Figure 9-11

quotient graph is given. Note that the vertex of degree one has a
current of order three going into it; the two singular edges carry
currents of order two, and the vertex of degree three satisfies the
KCL: (23)(12)(132) = (1)(2)(3) = e.
        There is one coset, $\Omega = S_3$. We have:

$$e: \quad (23),(12),(132),(123)$$
$$(12): \quad (132),e,(23),(13)$$
$$(23): \quad e,(123),(13),(12)$$
$$(13): \quad (123),(132),(12),(23)$$
$$(123): \quad (13),(23),e,(132)$$
$$(132): \quad (12),(13),(123),e \ .$$

The orbits are:

e-(23)-(123)	(12)-(13)-(23)
e-(12)-(23)	(12)-(132)-(13)
e-(132)-(12)	(23)-(13)-(123)
e-(123)-(132)	(13)-(132)-(123) .

        Figure 9-12 shows this imbedding, which agrees with Figure 9-10,
except for labeling.

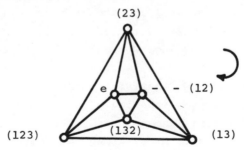

(23)

e   (12)

(123)   (132)   (13)

Figure 9-12

Example 2: Consider the quadrilateral imbedding of $K_{4,4}$ (for the
quaternions) given in Figure 7-3. We find the quotient graph for
this imbedding and choice of groups. Looking at the dual, we find
that e, y, x^2, and x^2y have the same boundary: x,x^2y,x^3,y.

Furthermore, $\Omega = \{e,y,x^2,x^2y\}$ is a subgroup of Q, since $x^2 = y^2$ and $x^2y = y^3$, with $y^4 = e$. Also, $\Omega x = \{x,x^3y,x^3,xy\}$ is a coset of Q with respect to Ω, and each of these regions (in the dual) has boundary: x^2y,x,y,x^3. Hence this imbedding of $K_{4,4}$ (for Q) is of <u>index two</u> ($[Q:\Omega] = |Q|/|\Omega| = 2$.) We "mod out" regions (in the dual) with identical boundaries, and form the quotient graph, in its quotient manifold S_1, as indicated in Figure 9-13. To make the appropriate identification, we use the general relationship:

$$(X,g)^{-1} = (Xg,g^{-1}),$$

to match the edge g in the boundary of the region for coset X. The computations for this situation are:

$$(\Omega,x)^{-1} = (\Omega x,x^3)$$
$$(\Omega,x^2y)^{-1} = (\Omega,y)$$
$$(\Omega,x^3)^{-1} = (\Omega x,x)$$
$$(\Omega x,y)^{-1} = (\Omega x,x^2y).$$

Example 3: Finally, we find a quadrilateral imbedding of K_{21} in S_{43} (with $r = r_4 = 105$). It is an easy exercise to check that this imbedding is compatible with the euler formula. Let us hope for the best, and try for an index one imbedding with $\Gamma = Z_{21}$. Then we seek a quotient manifold with $r = r_{20} = 1$. This requires a quotient graph K with 20 directed, or 10 undirected, edges. For a quadrilateral imbedding, we would like K to be regular, of degree 4. A likely candidate is $K = K_5$. Now we must make an assignment $\lambda: K^* \to Z_{21} - 0$, satisfying the KCL; then we must make a selection $(p_1,p_2,\ldots,p_5)$ inducing a single orbit, of length 20, for K_5. We find the assignment, using the fact that K_5 is eulerian; the selection is made, using the fact that $\gamma_M(K_5) = 3$ (i.e. S_3 will be the quotient manifold), with $r = r_{20} = 1$; see Figure 9-14.

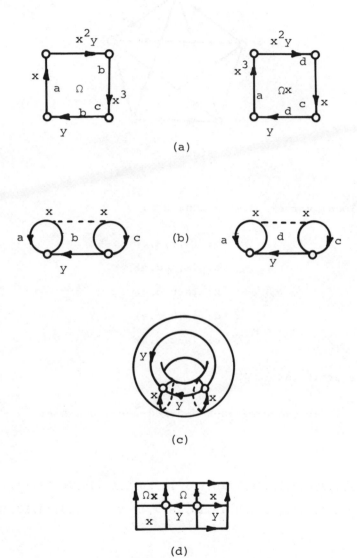

(a)

(b)

(c)

(d)

Figure 9-13

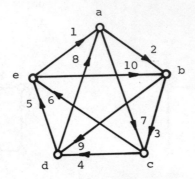

Figure 9-14

The local vertex permutations are:

$$a: \ (b,c,d,e)$$
$$b: \ (a,c,d,e)$$
$$c: \ (a,b,d,e)$$
$$d: \ (a,b,c,e)$$
$$e: \ (a,c,d,b)$$

The single orbit is:

a-b-c-d-e-b-a-c-b-d-c-e-d-a-e-c-a-d-b-e,

giving rise to:

0: 1,2,3,4,5,10,19,7,18,9,17,6,16,8,20,15,14,13,12,11

1: 2,3,4,5,6,11,20,8,19,10,18,7,17,9,0,16,15,14,13,12

.
.
.

20: 0,1,2,3,4,9,18,6,17,8,16,5,15,7,19,14,13,12,11,10 ,

which in turn gives $r = r_4 = 105$, for K_{21} on S_{43}, using Edmonds' permutation technique.

Although this imbedding may be of little general interest, it does give further indication of the economy of effort possible by seeking quotient graphs in quotient manifolds, rather than graphs in manifolds. Moreover, minimal imbeddings have been found by first finding non-minimal imbeddings of related graphs. For example, Youngs [76] imbeds K_{12s+7} on $S_{12s^2+13s+3}$, to find K_{12s+10} on $S_{12s^2+13s+4}$.

9-5. The General Theory

As we have seen, the standard proof-technique for establishing a genus or maximum genus formula is to first use the Euler formula

$$p - q + r = 2 - 2k,$$

describing a 2-cell imbedding of the graph G in the surface S_k, to find a bound for the parameter under study. For $\gamma(G)$, we get a lower bound (see, for example, Corollaries 6-14 and 6-15); for $\gamma_M(G)$ we get an upper bound (Theorem 6-24). The second step is to construct an imbedding of G attaining the bound obtained. Inductive constructions are particularly useful for graphs which can be defined as repeated cartesian products (as in the proofs of Theorems 6-30 and 7-12). A local vertex permutation scheme, such as that discussed in Section 6-6, is frequently helpful (as in the proofs of Theorems 6-25, 6-31, and 6-33). Ad hoc techniques have occasionally been successful (see the proof of Theorem 6-34).

But the most powerful and efficient technique (when it applies, and this is reasonably often!) is the method of quotient graphs and quotient manifolds. In the preceding sections of this chapter, we have discussed this theory, as introduced by Gustin and developed by Youngs, and its application to triangular imbeddings of complete graphs. We have also hinted at a generalization of this theory, and to this we now devote our attention.

The material in this section is based on work by Jacques [32].

Given a finite group Γ with a set Δ of generators for Γ, recall that the Cayley color graph $D_\Delta(\Gamma)$ has vertex set Γ, with (g,g') a directed edge -- labeled with generator δ_i -- if and only if $g' = g\delta_i$. We assume that, if $\delta_i \in \Delta$, $\delta_i^{-1} \notin \Delta$, unless δ_i has order 2. In this latter case, the two directed edges $(g,g\delta_i)$

and $(g\delta_i, g)$ are represented as a single undirected edge $[g, g\delta_i]$, labeled with δ_i. The graph obtained by deleting all labels and arrows (directions) from the edges of $D_\Delta(\Gamma)$ we call the <u>Cayley graph</u>, $G_\Delta(\Gamma)$. This graph has all edges of the form $[g, g\delta_i]$, for $g \in \Gamma$ and $\delta_i \in \Delta$.

As examples of graphs which are Cayley graphs, we have Q_n for the elementary 2-group $(Z_2)^n$, $K_{4,4} \times Q_n$ (where "x" denotes the cartesian product of two graphs) for the hamiltonian group $Q \times (Z_2)^n$ (where Q denotes the quaternions and here "x" indicates the direct product of two groups), K_n for the cyclic group Z_n, $K_{2n,2n}$ for Z_{4n}, $K_{n,n,n}$ for Z_{3n}, and the graph Π_n of Jacques [32] for the symmetric group S_n. The genus is known for each of the above families of graphs, and in each case the computation is facilitated by a knowledge of the theory we are about to present.

Consider a Cayley graph $G_\Delta(\Gamma)$ 2-cell imbedded in a closed orientable 2-manifold M. We study the imbedding of the Cayley color graph, $D_\Delta(\Gamma)$, thus determined. Let $\Delta^{-1} = \{\delta_i^{-1} | \delta_i \in \Delta\}$, and form $\Delta^* = \Delta \cup \Delta^{-1}$. The elements of Δ^* are called <u>currents</u>. The imbedding is characterized (see Section 6-6) by giving, at each $g \in \Gamma$, the cyclic permutation σ_g of $g\Delta^*$ determined by the orientation. For example, see Figure 9-15. Let Ω be a subgroup of Γ such that if $\Omega g = \Omega g'$, then $\sigma_g^* = \sigma_{g'}^*$, where σ_h^* is the

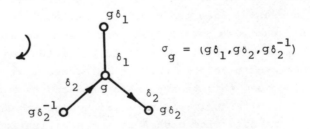

$$\sigma_g = (g\delta_1, g\delta_2, g\delta_2^{-1})$$

Figure 9-15

cyclic permutation of Δ^* induced by the action of σ_h on $h\Delta^*$; such a subgroup always exists, since we can take $\Omega = \{e\}$, where

e is the identity of Γ. It is to our advantage, however, to
choose Ω as large as possible. In the terminology of Jacques, Γ
and Ω determine a <u>quotient</u> <u>constellation</u> C' for the <u>constella-</u>
<u>tion</u> C = (D$_\Delta$(Γ) in M). The quotient constellation is an imbed-
ding of the Schreier coset graph (see Section 4-3) for Ω in Γ,
the imbedding being determined by the collection {σ$_h^*$}, taken over
any set {h} of right coset representatives of Ω in Γ. The
<u>reduced</u> <u>constellation</u> (C')* is the dual of the quotient constella-
tion. This is a 2-cell imbedding of a pseudograph K (with each
edge directed and labeled with the current of its dual edge; see
Figure 9-16) in a closed orientable 2-manifold, called by Youngs [75]
the <u>quotient</u> <u>graph</u> and <u>quotient</u> <u>manifold</u> respectively, for D$_\Delta$(Γ)
and Ω. Youngs, as we have seen in Section 9-2, obtains this struc-
ture by a different process: he first takes the dual of D$_\Delta$(Γ) in
M, and then "mods out" faces with identically labeled boundaries in
accordance with the subgroup Ω. Jacques' approach is consistent
with that of Gustin [26] and has the advantage of applying also to
<u>irregular</u> imbeddings (i.e. r ≠ r$_k$, for any k).

in C': $\Omega g'$ $h \in g^{-1}\Omega g'$ in (C')*: $\Omega g'$ h Ωg

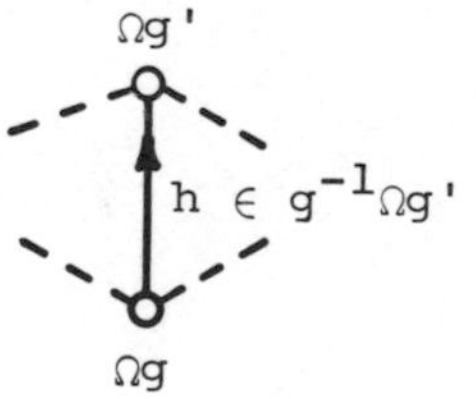

Ωg

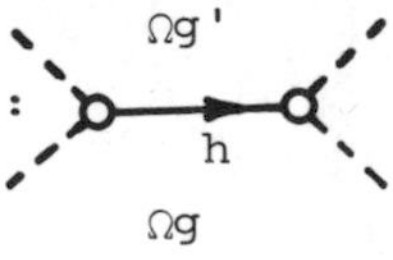

Figure 9-16

Define (after Jacques) a <u>brin</u> to be an ordered pair (c, δ*),
where c is a vertex in C, C' or (C')*, and δ* ∈ Δ*. Then we
have:

<u>Thm. 9-10.</u>   The reduced constellation (quotient graph in its quo-
                tient manifold, or M(Γ/Ω)) satisfies the following

five properties:

1) each brin carries a current from $\Delta*$
2) two opposing brins $x = (c, \delta*)$ and $x^{-1} = (c', \delta*^{-1})$
    (see Figure 9-17) carry inverse currents (if
    $x = x^{-1}$, the current must be of order 2; in this
    case the brin appears as in Figure 9-18).

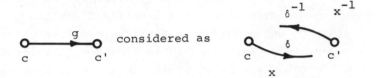

considered as

Figure 9-17

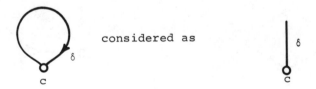

considered as

Figure 9-18

3) the regions are in one-to-one correspondence with
    the right cosets of Ω in Γ. (The index of Ω in
    Γ is called the <u>index of the imbedding</u>.)
4) the currents appearing in a region boundary are in
    one-to-one correspondence with $\Delta*$
5) if a brin x appears in the boundary of a region
    associated with Ωg and its opposite brin x^{-1} in
    the boundary of a region associated with $\Omega g'$, then

the current carried by x is in the set $g^{-1}\Omega g'$

Proof: The properties follow directly from the definitions and the construction of (C')*.

What is important for imbedding problems is the converse:

Thm. 9-11. A reduced constellation $M(\Gamma/\Omega)$ for $D_\Delta(\Gamma)$ and Ω
(i.e. a pseudograph 2-cell imbedded, with edges and re-
gions labeled with elements of Δ^* and right cosets of
Ω in Γ respectively) satisfying the five properties
above determines a 2-cell imbedding C of $D_\Delta(\Gamma)$ in
M such that (C')* = $M(\Gamma/\Omega)$.

Proof: For each $g \in \Gamma$, define $\sigma_g = (g\delta_1, g\delta_2, \ldots, g\delta_s)$,
where $\delta_1, \delta_2, \ldots, \delta_s$ is the oriented boundary of the
region $\Omega g'$ in $M(\Gamma/\Omega)$, and $g \in \Omega g'$. By property 3),
σ_g is well-defined; by properties 1) and 4), σ_g is a
permutation of $g\Delta^*$. By Edmonds' algorithm (Theorem
6-33), the collection $\{\sigma_g\}$ determines a 2-cell imbedding C of $D_\Delta(\Gamma)$ in M. Moreover, by properties 2)
and 5), the dual of $M(\Gamma/\Omega)$ is a Schreier coset graph
C' for C, so that (C')* = $M(\Gamma/\Omega)$.

In short, the region boundaries for $M(\Gamma/\Omega)$ determine the
Edmonds' permutation scheme for the Schreier coset graph imbedded in
the quotient constellation, and hence for $D_\Delta(\Gamma)$ (and thus $G_\Delta(\Gamma)$)
in M.

But we can learn much more from $M(\Gamma/\Omega)$. First we make the
following:

Def. 9-12. Let ξ be a vertex of $M(\Gamma/\Omega)$, with π the product
of currents (from Δ^*) directed away from ξ, in the
order given by the orientation. The order of π in Γ
is called the valence of ξ.

Remark. If Γ is non-abelian, π may not be uniquely determined;
nevertheless, if π and π' are two products at vertex
ξ, then π and π' are conjugate elements of Γ and
thus have the same order, so that the valence of ξ _is_
well-defined.

We now state:

Thm. 9-13. A vertex of degree k and valence ν in $M(\Gamma/\Omega)$ de-
            termines $\frac{|\Omega|}{\nu}$ regions of length $k\nu$ in the imbedding
            of $D_\Delta(\Gamma)$ in M.

            Proof: Let ξ be a vertex in $M(\Gamma/\Omega) = (C')*$, of
            degree k and valence ν. Thus, in C', ξ corres-
            ponds to a region of length k for which the product
            $\pi = \delta_1\delta_2\ldots\delta_k$ of currents in the oriented boundary
            (from, say, coset Ω_g to coset Ω_g) is of order ν.
            In C, the walk determined by π is a portion of a re-
            gion boundary, from g' to g", where g' and g"
            are both in Ωg. This region boundary continues with
            another walk determined by π (since $\sigma_{g'} = \sigma_{g''}$) and
            only concludes at g', for $\pi^\nu = e$. Thus this region
            in C is of length $k\nu$. In this region boundary we
            find exactly ν brins of the form (h, δ_1), with
            $h \in \Omega g$; hence ξ determines $\frac{|\Omega|}{\nu}$ regions of length $k\nu$
            in $C = (D_\Delta(\Gamma)$ in M).

    Thus imbeddings can be studied in terms of possibly much
simpler combinatorial structures: the reduced constellations of
Jacques or, equivalently, the quotient graphs and quotient manifolds
of Youngs.

9-6. Applications to Known Imbeddings

    To illustrate the power and beauty of the theory, we present
four examples, each giving a _new_ proof of a genus formula appearing
in the literature. We then give a fifth example, illustrating what
Youngs called the theory of vortices.

Application 1: Take $\Gamma = Z_8$, $\Omega = Z_2 = \{0,4\}$, and $\Delta = \{1,3\}$.
                Then $M(Z_8/Z_2)$ in Figure 9-19 determines an imbed-
                ding of $K_{4,4} = G_\Delta(Z_8)$ in S_1, with $r = r_4 =$
                $4\frac{|\Omega|}{\nu} = 8$. This is an index four imbedding. The
                generalization to $K_{2n,2n}$ is fairly easy to obtain,
                with $\Gamma = Z_{4n}$ and $\Omega = Z_n$; $M(Z_{4n}/Z_n)$ is a

multigraph of order $2n$ (with edges appropriately directed and labeled) in S_{n-1}, giving
$K_{2n,2n}$ quadrilaterally imbedded in $S_{(n-1)^2}$.
(This imbedding appears in Ringel [53].)

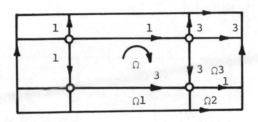

Figure 9-19

Application 2: Take $\Gamma = Z_9$, $\Omega = Z_3 = \{0,3,6\}$, and $\Delta = \{1,2,4\}$.
Then $M(Z_9/Z_3)$ in Figure 9-20 gives $K_{3,3,3} = G_\Delta(Z_9)$ in S_1, with $r = r_3 = \frac{6\lfloor\Omega\rfloor}{\nu} = 18$. This is
an index three imbedding. The generalization to
$K_{2m+1,2m+1,2m+1}$, with $\Gamma = Z_{6m+3}$ and $\Omega = Z_{2m+1}$,
has S_m as quotient manifold, and produces a triangular imbedding of $K_{2m+1,2m+1,2m+1}$ in $S_{m(2m-1)}$;
this imbedding appears in [71].

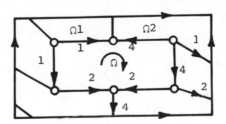

Figure 9-20

<u>Application 3</u>: Take $\Gamma = Q \times Z_m$,   where m is odd, $\Omega = Z_4 \times Z_m$,
and $\Delta = \{s, ta\}$, where Q is generated by s and
t, and Z_m is generated by a.   Then

$M(Q \times Z_m / Z_4 \times Z_m)$ in Figure 9-21 gives $G_\Delta(Q \times Z_m)$
in S_1,   with $r = r_4 = 2\frac{|\Omega|}{\nu} = 8m$.   (This index
two imbedding was first found by Himelwright [21];
it shows that the hamiltonian group $Q \times Z_m$ is
toroidal.)

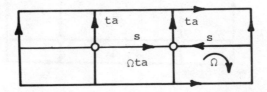

Figure 9-21

<u>Application 4</u>: Consider $\Gamma = \Omega = S_n$ (n > 3),   and $\Delta = \{s, t\}$,   with
$s = (1\ 2\ \dots\ n)$ and $t = (1\ 2)$.   Then $M(S_n/S_n)$ in
Figure 9-22 gives $G_\Delta(S_n)$ in S_k,   where k =
$1 + \frac{(n-2)!}{4}(n^2 - 5n + 2)$,   with $r_n = \frac{n!}{n} = (n-1)!$ and
$r_{2n-2} = \frac{n!}{n-1}$.   The brin carrying current t represents a loop.   (This index one imbedding appears in
[70], and is conjectured to give the genus of the
symmetric group S_n,   for n even.)

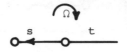

Figure 9-22

Application 5: Take $\Gamma = \Omega = Z_7$, and $\Delta = \{1,2,3\}$. Then $M(Z_7/Z_7)$ in Figure 9-23 gives $G_\Delta(Z_7) = K_7$ in S_3,

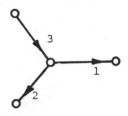

Figure 9-23

with $r_3 = 7$ and $r_7 = 3$. (This index one imbedding solves the regular part of the problem, for finding $\gamma(K_{10})$; see Section 9-3.)

9-7.   New Applications

Further connections between the theory and imbedding problems can be explored. For example:

Thm. 9-14.   If K is a quotient graph for an imbedding of a graph G of index one or two, then $\gamma_M(K) = \left\lceil \frac{\beta(K)}{2} \right\rceil$.

Proof: This is a direct consequence of Theorem 6-24.

Every Cayley graph has an index one imbedding; in fact $G_\Delta(\Gamma)$ has $(|\Delta^*|-1)!$ index one imbeddings. When are these minimal? A partial answer is provided in:

Thm. 9-15.   Let the finite group Γ be generated by $\Delta = \{\delta_1,\ldots,$
$\delta_n\}$, where each δ_i is of order $n \geq 3$ and $\prod\limits_{i=1}^{n} \delta_i = e$.
Then $G_\Delta(\Gamma)$ has an index one imbedding, with $r = r_n = 2|\Gamma|$, on S_k, where $k = 1 + \frac{|\Gamma|}{2}(n-3)$. If $n = 3$, or if $n = 4$ and no proper subset of Δ contains a redundant generator, then $\gamma(G_\Delta(\Gamma)) = k$.

Cor. 9-16. Let $\Gamma = (Z_n)^{n-1}$, $n \geq 3$, with $\Delta = \{\varepsilon_1, \ldots, \varepsilon_{n-1}, (\prod_{i=1}^{n-1} \varepsilon_i)^{-1}\}$, where ε_i is the (n-1)-tuple having 1 in the ith position and 0 elsewhere. Then

$$\gamma(G_\Delta(\Gamma)) \leq 1 + \frac{n^{n-1}(n-3)}{2}; \quad \text{equality holds for} \quad n = 3,4.$$

Cor. 9-17. Let $\Gamma = (3,3|3,3)$ in the notation of Coxeter and Moser [15] (i.e. Γ is the non-abelian group of order 27, generated by s,t and subject to the relations $s^3 = t^3 = (st)^3 = (s^{-1}t)^3 = e$); let $\Delta = \{s,t,(st)^2\}$; then $\gamma(G_\Delta(\Gamma)) = 1$. Moreover, Γ is a toroidal group.

Cor. 9-18. Let $\Gamma = A_4$, the alternating group of degree four, with $\Delta = \{(1\ 2\ 3), (3\ 2\ 4), (1\ 3\ 4)\}$; then $\gamma(G_\Delta(A_4)) = 1$. (However, A_4 is not toroidal, as there exist planar Cayley graphs for A_4.)

In Section 7-2 we saw that we could find the genus of any abelian group $\Gamma = Z_{m_1} \times Z_{m_2} \times \ldots \times Z_{m_r}$, where m_r is even (assuming m_i divides m_{i-1}, $i = 2, \ldots, r$.) The argument would use the fact that $C_{m_1} \times C_{m_2} \times \ldots \times C_{m_r} = G_\Delta(\Gamma)$, for the standard presentation for Γ, and Theorem 7-12. The techniques employed fail if even one order m_i is odd. The theory of quotient graphs and quotient manifolds provides a new approach to the genus problem for general cycles, and hence to the genus of an arbitrary finite abelian group.

For example, $M(Z_3 \times Z_{2n} \times Z_{2m}/Z_3 \times Z_n \times Z_{2m})$ in Figure 9-24, where Z_3, Z_{2n}, and Z_{2m} are generated by r, s, and t respectively, shows that

$$\gamma(C_3 \times C_{2n} \times C_{2m}) \leq 1 + 4mn;$$

we conjecture equality for this formula. Also, $M((Z_3)^n/(Z_3)^n)$ in Figure 9-25, where ε_i is a generator for the ith copy of Z_3, shows that

$$\gamma((C_3)^n) \leq 1 + (n-2)3^{n-1}.$$

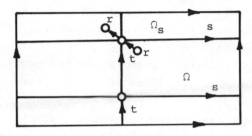

Figure 9-24

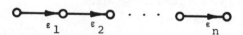

Figure 9-25

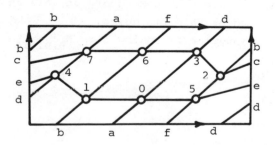

Figure 9-26

9-8. Problems

*9-1.) Show that seven "different" minimal imbeddings of K_5 are
compatible with the formula of Theorem 9-1. (For example,
$r_3 = 4$, $r_8 = 1$; $r_4 = 5$ give two "different" imbeddings of
K_5 on S_1, since the regions are distributed differently).
How many of the seven can you actually find? (Hint: not all
seven exist!)

9-2.) Find the quotient manifold and the quotient graph for the
quadrilateral imbedding of $K_{4,4}$ given (for $\Gamma = Z_8$) in
Figure 9-26.

*9-3.) There are ten different quotient graphs $S(\Gamma/\Omega)$ for
$K_{2,2,2,2}$ in S_1 with $\Omega \neq \{e\}$; i.e. ten different ordered
pairs (Γ, Ω) for $K_{2,2,2,2}$. Find these ten ordered pairs
and then obtain an $M(\Gamma/\Omega)$ for each non-trivial index.

9-4.) Use the theory of quotient graphs and quotient manifolds to
find a quadrilateral imbedding of K_5.

9-5.) Use the theory of quotient graphs and quotient manifolds to
find a hexagonal imbedding $(r = r_6)$ for $K_{3,3}$.

*9-6.) Generalize example 3 of Section 9-4, to find quadrilateral
imbeddings of $K_{4n(4n+1)+1}$, where $4n+1$ is prime. Is this
an infinite class of graphs?

9-7.) Prove Theorem 9-15.

9-8.) Find $\gamma(Q_n)$ in three ways: (1) by a construction, such as
that employed in the proof of Theorem 7-12; (2) by selecting
an appropriate $(p_1, p_2, \ldots, p_{2^n})$ and using Edmonds'
algorithm (see Theorem 6-33) to show that every orbit has
length 4; (3) by finding an appropriate $M(\Gamma/\Omega)$.

9-9.) We see from Theorem 6-25 that K_5 has a 2-cell imbedding
in S_3. Show that the theory of this chapter is of no
aid in finding this imbedding.

**9-10.) Find a triangular imbedding for $K_{n,n,n,n}$ in a suitable
surface.

References

1.) S. Anderson, Graph Theory and Finite Combinatorics, Markham, Chicago, 1970.

2.) G. Atneosen, On the Embeddability of Compacta in n-Books: Intrinsic and Extrinsic Properties, Ph.D. Thesis, Michigan State University, East Lansing, Mich., 1968.

3.) J. Battle, F. Harary, Y. Kodama, and J. W. T. Youngs, Additivity of the genus of a graph, Bull. Amer. Math. Soc. 68 (1962), 565-568.

4.) M. Behzad and G. Chartrand, An Introduction to the Theory of Graphs, Allyn and Bacon, Boston, 1971.

5.) L. W. Beineke, Complete bipartite graphs: decomposition into planar subgraphs, Chapter 7 in A Seminar in Graph Theory (F. Harary, ed.), Holt, Rinehart and Winston, New York, 1967.

6.) L. W. Beineke, The decomposition of complete graphs into planar subgraphs, Chapter 4 in Graph Theory and Theoretical Physics (F. Harary, ed.), Academic Press, London, 1967.

7.) L. W. Beineke and F. Harary, The genus of the n-cube, Can. J. Math. 17 (1965), 494-496.

8.) G. Birkhoff and S. MacLane, Algebra, MacMillan, New York, 1967.

9.) H. R. Brahana, Regular maps on an anchor ring, Amer. J. Math. 48 (1926), 225-240.

10.) W. Burnside, Theory of Groups of Finite Order, Second Edition, Cambridge, 1911.

11.) A. Cayley, The theory of groups: a graphical representation, Amer. J. Math. 1 (1879), 174-176.

12.) G. Chartrand and D. C. Kay, A characterization of certain ptolemaic graphs, Canad. J. Math. 17 (1965), 342-346.

13.) G. Chartrand and H. V. Kronk, The point-arboricity of planar graphs, J. London Math. Soc. 44 (1969), 612-616.

14.) G. Chartrand, H. V. Kronk, and C. E. Wall, The point-arboricity of a graph, Israel J. Math. 6 (1968), 168-175.

15.) H. S. M. Coxeter and W. O. Moser, Generators and Relations for Discrete Groups, Springer-Verlag, Berlin, 1957.

16.)   A. K. Dewdney, The chromatic number of a class of pseudo-
       2-manifolds, Manuscripta Math. 6 (1972), 311-320.

17.)   G. A. Dirac, A theorem of R. L. Brooks and a conjecture of
       H. Hadwiger, Proc. London Math. Soc. (3) 7 (1957), 161-195.

18.)   G. A. Dirac and S. Schuster, A theorem of Kuratowski, Indag.
       Math. 16 (1954), 343-348.

19.)   R. A. Duke, The genus, regional number, and Betti number of a
       graph, Canad. J. Math. 18 (1966), 817-822.

20.)   W. Dyck, Gruppentheoretische Studien, Math. Ann. 20 (1882),
       1-45.

21.)   J. Edmonds, A combinatorial representation for polyhedral
       surfaces, Notices Amer. Math. Soc. 7 (1960), 646.

22.)   P. Franklin, A six colour problem, J. Math. Phys.,
       Massachusetts Inst. Technol. 13 (1934), 363-369.

23.)   M. Fréchet and K. Fan, Initiation to Combinatorial Topology,
       Prindle, Weber and Schmidt, Boston, Mass., 1967.

24.)   R. Frucht, Herstellung von Graphen mit vorgegebener
       abstrakten Gruppe, Compositio Math. 6 (1938), 239-250.

25.)   H. Grötzsch, Zur Theorie der diskreten Gebilde, VII, Ein
       Dreifarbensatz fur dreikreisfreie Netze auf der Kugel, Wiss.
       Z. Martin-Luther-Univ. Halle-Wittenberg, Math. Nat. Reihe 8
       (1958/59), 109-120.

26.)   W. Gustin, Orientable embedding of Cayley graphs, Bull. Amer.
       Math. Soc. 7 (1960), 646.

27.)   R. K. Guy, The decline and fall of Zarankiewicz's theorem,
       Proof Techniques in Graph Theory (F. Harary, ed.), Academic
       Press, New York, 1969.

28.)   F. Harary, Graph Theory, Addison-Wesley, Reading, Mass., 1969.

29.)   P. J. Heawood, Map-colour theorem, Quart. J. Math. 24 (1890),
       332-338.

30.)   L. Heffter, Uber das Problem der Nachbargebiete, Math. Ann.
       38 (1891), 477-508.

31.)   P. Himelwright, On the genus of Hamiltonian groups, Specialist
       Thesis, Western Michigan University, Kalamazoo, Mich., 1972.

32.)   A. Jacques, Constellations et Propriétés Algèbriques des
       Graphes Topologiques, Ph.D. Thesis, University of Paris, 1969.

33.)   M. Kleinert, Die Dicke des n-dimensionalen Würfel-Graphen,
       J. Combinatorial Theory 3 (1967), 10-15.

34.)   H. V. Kronk, An analogue to the Heawood map-coloring problem,
       J. London Math. Soc. 1 (Ser. 2), (1969), 550-552.

35.) H. V. Kronk and A. T. White, A 4-color theorem for Toroidal
     Graphs, Proc. Amer. Math. Soc. 34 (1972), 83-86.

36.) G. Kuratowski, Sur le Probleme des Courbes Gauches en
     Topologie, Fund. Math. 15 (1930), 271-283.

37.) H. Levinson, On the genera of graphs of group presentations,
     Ann. New York Acad. Sci. 175 (1970), 277-284.

38.) D. R. Lick and A. T. White, k-degenerate graphs, Canad. J.
     Math. 22 (1970, 1082-1096.

39.) D. R. Lick and A. T. White, The point partition numbers of
     closed 2-manifolds, J. London Math. Soc. 4 (1972), 577-583.

40.) Magnus, Karass, and Solitar, Combinatorial Group Theory, Inter-
     science Publishers, New York, 1966.

41.) W. Maschke, The representation of finite groups, Amer. J. Math.
     18 (1896), 156-194.

42.) W. S. Massey, Algebraic Topology; an Introduction, Harcourt,
     Brace, and World, 1967.

43.) J. Mayer, Le Problème des Régions Voisines sur les Surfaces
     Closes Orientables, J. Combinatorial Theory 6 (1969), 177-195.

44.) J. Mycielski, Sur le coloriage des graphs, Colloq. Math. 3
     (1955), 161-162.

45.) E. A. Nordhaus, R. D. Ringeisen, B. M. Stewart, and A. T.
     White, A Kuratowski-type theorem for the maximum genus of a
     graph, J. Combinatorial Theory B12 (1972), 260-267.

46.) E. A. Nordhaus, B. M. Stewart, and A. T. White, On the maximum
     genus of a graph, J. Combinatorial Theory B 11 (1971), 258-267.

47.) O. Ore, The Four-color Problem, Academic Press, New York,1967.

48.) O. Ore and G. J. Stemple, Numerical methods in the four color
     problem, Recent Progress in Combinatorics (W. T. Tutte, ed.),
     Academic Press, New York, 1969.

49.) W. Petroelje, Imbedding Graphs on Pseudosurfaces, Specialist
     Thesis, Western Michigan University, Kalamazoo, Mich., 1971.

50.) R. D. Ringeisen, Determining all compact orientable 2-manifolds
     upon which $K_{m,n}$ has 2-cell imbeddings, J. Combinatorial Theory
     B12 (1972), 101-104.

51.) R. D. Ringeisen, The Maximum Genus of a Graph, Ph.D. Thesis,
     Michigan State University, East Lansing, Mich., 1970.

52.) G. Ringel, Färbungsprobleme auf Flächen und Graphen,
     Deustscher Verlag, Berlin, 1959.

53.) G. Ringel, Das Geschlecht des Vollständigen Paaren Graphen,
     Abh. Math. Sem. Univ. Hamburg 28 (1965), 139-150.

54.) G. Ringel, Über Drei Kombinatorische Probleme am n-dimension-
      alen Würfel and Würfelgitter, Abh. Math. Sem. Univ. Hamburg
      20 (1965), 10-19.

55.) G. Ringel and J. W. T. Youngs, Das Geschlecht des Symmetrische
      Vollständige Drei-Farbaren Graphen, Comment. Math. Helv. 45
      (1970), 152-158.

56.) G. Ringel and J. W. T. Youngs, Solution of the Heawood map-
      coloring problem, Proc. Nat. Acad. Sci. U.S.A. 60 (1968),
      438-445.

57.) G. Ringel and J. W. T. Youngs, Solution of the Heawood map-
      coloring problem -- case 2, J. Combinatorial Theory 7 (1969),
      342-353.

58.) G. Ringel and J. W. T. Youngs, Solution of the Heawood map-
      coloring problem -- case 8, J. Combinatorial Theory 7 (1969),
      353-363.

59.) G. Ringel and J. W. T. Youngs, Solution of the Heawood map-
      coloring problem -- case 11, J. Combinatorial Theory 7 (1969),
      71-93.

60.) G. Sabidussi, The composition of graphs, Duke Math. J. 26
      (1959), 693-696.

61.) G. Sabidussi, Graph multiplication, Math. Z. 72 (1960),
      446-457.

62.) E. H. Spanier, Algebraic Topology, McGraw Hill, New York,
      1966.

63.) E. Steinitz, Polyeder und Raumeinteilungen, Enzykl. Math.
      Wiss. 3 (1922), 1-139.

64.) B. M. Stewart, Adventures Among the Toroids, B. M. Stewart,
      Okemos, Mich., 1970.

65.) C. M. Terry, L. R. Welch, and J. W. T. Youngs, The genus of
      K_{12s}, J. Combinatorial Theory 2 (1967), 43-60.

66.) C. M. Terry, L. R. Welch, and J. W. T. Youngs, Solution of the
      Heawood map-coloring problem -- case 4, J. Combinatorial
      Theory 8 (1970), 170-174.

67.) H. Tietze, Famous Problems of Mathematics, Graylock Press,
      New York, 1969.

68.) L. Weinberg, On the maximum order of the automorphism group
      of a planar triply connected graph, J. SIAM Appl. Math. 14
      (1966), 729-738.

69.) M. J. Wenninger, Polyhedron Models for the Classroom, National
      Council of Teachers of Mathematics, Washington, D.C., 1966.

70.)   A. T. White, On the genus of a group, _Trans_. _Amer_. _Math_. _Soc_.,
       to appear.

71.)   A. T. White, The genus of the complete tripartite graph
       $K_{mn,n,n}$, _J_. _Combinatorial_ _Theory_ 7 (1969), 283-285.

72.)   A. T. White, On the genus of the composition of two graphs,
       Pacific _J_. _Math_. 41 (1972), 275-279.

73.)   A. T. White, The genus of repeated cartesian products of bi-
       partite graphs, _Trans_. _Amer_. _Math_. _Soc_. 151 (1970), 393-404.

74.)   J. W. T. Youngs, "The Heawood Map-coloring Conjecture," Rand
       Corporation Memorandum RM - 4752 - PR, March 1966.

75.)   J. W. T. Youngs, "The Heawood Map Coloring Conjecture,"
       Chapter 12 in _Graph_ _Theory_ _and_ _Theoretical_ _Physics_ (F. Harary,
       ed.), Academic Press, London, 1967.

76.)   J. W. T. Youngs, The Heawood map-coloring problem -- cases
       1,7, and 10, _J_. _Combinatorial_ _Theory_ 8 (1970), 220-231.

77.)   J. W. T. Youngs, Minimal imbeddings and the genus of a graph,
       _J_. _Math_. _Mech_. 12 (1963), 303-315.

78.)   J. W. T. Youngs, "The Mystery of the Heawood Conjecture," in
       Graph _Theory_ _and_ _its_ _Applications_ (B. Harris, ed.), Academic
       Press, New York, 1970, 17-50.

79.)   J. W. T. Youngs, Solution of the Heawood map-coloring problem
       -- cases 3,5,6, and 9, _J_. _Combinatorial_ _Theory_ 8 (1970),
       175-219.

80.)   J. Zaks, The maximum genus of cartesian products of graphs,
       to appear.

Bibliography

a.) S. Anderson, Graph Theory and Finite Combinatorics, Markham,
     Chicago, 1970.
b.) B. Arnold, Intuitive Concepts in Elementary Topology,
     Prentice-Hall, Englewood Cliffs, N.J., 1962.
c.) M. Behzad and G. Chartrand, An Introduction to the Theory of
     Graphs, Allyn and Bacon, Boston, 1971.
d.) H. S. M. Coxeter and W. O. Moser, Generators and Relations for
     Discrete Groups, Springer-Verlag, Berlin, 1957.
e.) M. Fréchet and K. Fan, Initiation to Combinatorial Topology,
     Prindle, Weber and Schmidt, Boston, Mass., 1967.
f.) I. Grossman and W. Magnus, Groups and Their Graphs, Random
     House, New York, 1964.
g.) F. Harary, Graph Theory, Addison-Wesley, Reading, Mass., 1969.
h.) W. Magnus, A. Karrass, and D. Solitar, Combinatorial Group
     Theory, Interscience Publishers, New York, 1956.
i.) W. S. Massey, Algebraic Topology; an Introduction, Harcourt,
     Brace, and World, New York, 1967.
j.) G. Ringel, Färbungsprobleme auf Flächen und Graphen,
     Deutscher Verlag, Berlin, 1959.
k.) H. Tietze, Famous Problems of Mathematics (Chapters IV and
     XI), Graylock Press, New York, 1965.
l.) J. W. T. Youngs, "The Heawood Map Coloring Conjecture," in
     Graph Theory and Theoretical Physics, Academic Press, London,
     1967, 313-354.
m.) J. W. T. Youngs, "The Heawood Map Coloring Conjecture," Rand
     Corporation Memorandum RM-4752-PR, March 1966.
n.) J. W. T. Youngs, "The Mystery of the Heawood Conjecture," in
     Graph Theory and its Applications (B, Harris, ed.), Academic
     Press, New York, 1970, 17-50.

Graphs		Groups		Surfaces	
G	5	$\underline{G(G)}$	15		
		$G_1(G)$	19		
		$G^*(G)$	19		
$G_P(\Gamma)$	69	Γ	22		
$G_\Delta(\Gamma)$	69				
$D_P(\Gamma)$	22				
$\underline{D_\Delta(\Gamma)}$	22	$G(D_\Delta)(\Gamma))$	24		
		$\underline{\Omega(S_k)}$	3	S_k	37
				$\widetilde{S}_k$	88
				M_n	88
				S'	63
$\underline{\gamma(G)}$	53	$\underline{\gamma(\Gamma)}$	69	$\gamma(S_k)$	37
$S(\Gamma/\Omega)$	108	Γ/Ω	27	$M(\Gamma/\Omega)$	107
$\chi(G)$	82	$\chi(\Gamma)$	101	$\underline{\chi(S_k)}$	82
				$\chi(\widetilde{S}_k)$	88
				$\chi(M_n)$	88
$\chi'(G)$	63			$\chi'(S')$	63
$\rho_k(G)$	98	$\rho_k(\Gamma)$	101	$\rho_k(M_n)$	98
$G_1 \cong G_2$	9	$\Gamma_1 \cong \Gamma_2$	14		
$G_1 = G_2$	9	$\Gamma_1 \equiv \Gamma_2$	14		
$G_1 + G_2$	10	$\Gamma_1 + \Gamma_2$	16		
$G_1 \times G_2$	10	$\Gamma_1 \times \Gamma_2$	16,29		
$G_1[G_2]$	10	$\Gamma_1[\Gamma_2]$	16		
		X	107	$[X]$	107
				$[X^*]$	107
P_i	61	σ_g	120	$K \subset L$	107
K_n	10	S_n	16		
W_m	49	D_n	16		
C_n	10	Z_n	16		
Q_n	11	$(Z_2)^n$	74		
$K_{m,n}$	11	Q	32		

- -

Graphs		Groups		Surfaces	
$V(G)$	5	e	107	r_i	42
$E(G)$	5	Δ	119	r	38
$\delta(G)$	9	Δ^*	120	E^n	35
$\Delta(G)$	9	δ_i	119	B^n	35
$g(G)$	94	P	22	$\overset{\circ}{D}$	34
$a(G)$	97	A_n	16	S^2	39

Note: The number following the symbol gives the first page on which the symbol appears. In the top half of the list, corresponding symbols are displayed alongside one another. The six distinguished symbols correspond to the six relationships depicted in Figure 0-1.